CABLE TELEVISION TECHNOLOGY

CABLE TELEVISION TECHNOLOGY

J. N. SLATER, T.Eng.(CEI), AMIERE
Engineering Secretariat
Independent Broadcasting Authority, Winchester

ELLIS HORWOOD LIMITED
Publishers · Chichester

Halsted Press: a division of
JOHN WILEY & SONS
New York · Chichester · Brisbane · Toronto

First published in 1988 by
ELLIS HORWOOD LIMITED
Market Cross House, Cooper Street,
Chichester, West Sussex, PO19 1EB, England
The publisher's colophon is reproduced from James Gillison's drawing of the ancient Market Cross, Chichester.

Distributors:

Australia and New Zealand:
JACARANDA WILEY LIMITED
GPO Box 859, Brisbane, Queensland 4001, Australia

Canada:
JOHN WILEY & SONS CANADA LIMITED
22 Worcester Road, Rexdale, Ontario, Canada

Europe and Africa:
JOHN WILEY & SONS LIMITED
Baffins Lane, Chichester, West Sussex, England

North and South America and the rest of the world:
Halsted Press: a division of
JOHN WILEY & SONS
605 Third Avenue, New York, NY 10158, USA

South-East Asia
JOHN WILEY & SONS (SEA) PTE LIMITED
37 Jalan Pemimpin # 05–04
Block B, Union Industrial Building, Singapore 2057

Indian Subcontinent
WILEY EASTERN LIMITED
4835/24 Ansari Road
Daryaganj, New Delhi 110002, India

© 1988 J. N. Slater/Ellis Horwood Limited

British Library Cataloguing in Publication Data
Slater, J.N.
Cable television technology. —
(Ellis Horwood series in electrical engineering).
1. Community antenna television — Apparatus and supplies
I. Title
621.388'35   TK6675

Library of Congress Card No. 87–36674

ISBN 0–7458–0108–0 (Ellis Horwood Limited)
ISBN 0–470–21064–8 (Halsted Press)

Phototypeset in Times by Ellis Horwood Limited
Printed in Great Britain by Butler & Tanner, Frome, Somerset

Table of Contents

Introduction

Although cable television has been in existence for almost as many years as broadcast television, in various forms, the recent enthusiasm which the UK Government has shown for the information-technology related aspects of cable has brought cable systems into the spotlight again, and in particular has concentrated attention on the many exciting new facilities that the so-called 'new-technologies' can bring to cable systems.

As with all technical developments, the reporting of cable television in the popular press is subject to all sorts of misunderstandings, inaccuracies and over-simplifications, so that it is difficult for the lay reader to obtain a balanced picture of what is actually happening, and even television engineers can be confused about the real impact that the various technological changes can bring. Cable stories in the press range from that of a utopian society where our every need is fed along an optical light-pipe — and it is amazing how frequently our need is expected to be 30 channels of television — to scary tales of the upheaval that the installation of cable systems can cause, usually illustrated by American-originated photographs of huge lorries mounted with enormous circular-saw blades that are chewing up our roads. Even in the popular technical journals, lasers and fibre-optics are generally considered to be a 'must' for all new systems, and the well-proven coaxial systems that have been developed and used over many years are consigned to the technological dustbin, a situation which is far from the truth.

This book brings together an enormous amount of information about all aspects of cable technology and attempts to give a balanced view of the likely ways in which the expansion of cable will affect us in the future. It contains a good deal of information for the lay reader who is seeking to grasp the basic facts about cable television engineering, and this is presented in a thoroughly readable style, but in addition it is a unique source of technical reference to the television engineer who is striving to keep up with the latest developments in the technology of cable distribution.

A complete overview of the various possibilities that cable systems can provide is followed by chapters on the design of systems, with information about the standards of performance and construction that should be expected, and various techniques for achieving satisfactory systems are discussed.

The pros and cons of fibre-optic and conventional metal cables are considered in detail, and the latest information about the cost-effectiveness of different combinations of cable in different systems is provided.

Cable television will almost certainly rely on the complementary techniques of satellite distribution to provide its systems with a wide range of programmes from all parts of the world, and the technical features of suitable satellite reception terminals are considered in some detail together with an assessment of the quality of reception that can be expected from various types of receiving equipment.

No study of cable television systems would be complete without a look at the interactive services that cable systems are uniquely equipped to provide. This book explains about the various services that could be provided, from home banking to teleshopping, from 'on-line' gambling to the downloading of computer software, and tries to give the reader a 'feel' for some of the exciting new possibilities that are yet to become reality.

New technologies give rise to new legislation, and the cable industry has always been subject to various restrictions. The role of the new legislatory bodies is discussed, together with the practical implications of the various rules and regulations on the cable operator and on the viewer who is keen to stay on the right side of the law.

Although this is not a book aimed at the professional cable engineer, the author has tried to include enough information to make the book a useful reference manual for the television service engineer who is not a cable television specialist, but at the same time providing a readable source of information for the engineering student who needs a clear understanding of the subject, as well as an interesting and easily assimilated overview of cable television for the less technical reader who merely wishes to skim through.

The Vermeer CRC31 concrete cutter in action for Balfour Kilpatrick in the streets of Croydon in the early stages of a project which will supply 115 000 homes in the area with a cable television service. Photo: Cable Television Engineering.

1

Cable television — a whole range of different systems

Cable television distribution has been used for almost 50 years, and initially came into existence as a way of providing television pictures and sound to people living in areas where satisfactory signals could not be received off-air. This was because the broadcasting authorities had not yet managed to provide enough transmitting stations for all those communities that wanted to receive the new wonder of television. Nowadays, when the virtues of private enterprise are being regularly trumpeted by a government that is often not prepared to provide the finance for major capital projects, it is instructive to look backwards to see how self-sufficient the founders of the early wired distribution systems were. Radio and electrical dealers who were keen to expand into the newly arrived television business rapidly realised that they would not be able to sell receivers to customers who could not pick up satisfactory signals, and so many dealers experimented by wandering over nearby hillsides, trying to find a spot where good pictures could be received with suitable aerials. These pioneers then proceeded to tackle the problems of buying a small piece of land on which to put the receiving equipment, obtaining permission to run cable down the hillsides and along the streets of many a small town, putting wires into the home of every would-be viewer who could be persuaded to subscribe. Even now, when countries like the United Kingdom have television coverage approaching almost 100% of the population, there are large numbers of very small communities which are geographically isolated or hidden away in hollows where they are screened from any incoming television signals, and these areas have to rely on signals being fed to them by wire (Fig. 1.1).

In some areas this simple type of system was developed by entrepreneurial dealers who discovered that at their aerial vantage points they could receive television programmes from other, more distant transmitters that were intended to provide services for other areas. Generally speaking they could provide one or more extra programmes from the regional independent television companies around the country, although in some parts, notably in the border country of Wales, they could also offer their customers a choice of English or Welsh BBC transmissions. Although

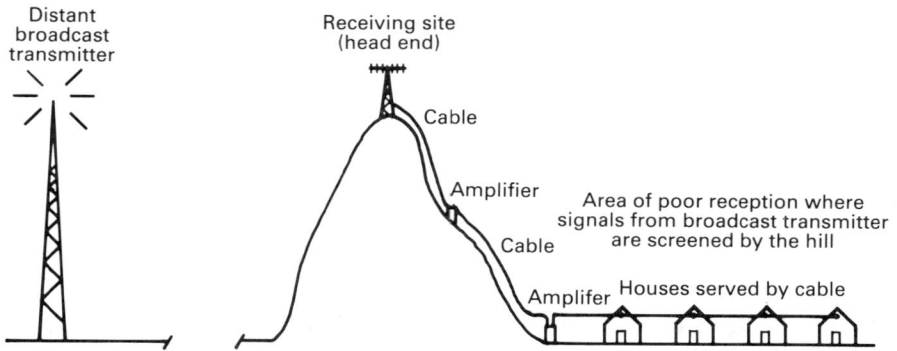

Fig. 1.1 — Simple cable service providing television signals to screened area.

this particular facility proved very popular, it turned out to be something of a political hot-potato, and regulations were introduced to restrict such cable services from supplying more than one extra programme, and they were generally not allowed to provide locally inserted material. In spite of this unhelpful regulatory climate, cable services remained popular in a few areas even after the transmitting authorities built local transmitters, because the cable operators could provide a wider choice of programmes than the local transmitters did. It is these systems that have become the forerunners of the sophisticated cable services that will develop and expand in the next few years.

The wired networks of the future will be able to receive signals not only from many different parts of this country, but will also be able to pick up and and re-distribute signals from all around the world, thanks to the marvels of satellite technology (Fig. 1.2). Locally produced services of news and information will be available in audio-visual and textual forms at all times of the day and night, and the unique two-way capabilities of modern cable systems will be used to enable the viewer to 'talk-back' to the programme provider, and to make use of a wide range of hitherto unexplored services such as teleshopping and home banking. We shall be looking at these different types of service in more detail later in the book, but it is perhaps worth mentioning at this point that most of these services can be provided by almost any type of well-engineered cable network, and the particular technology used is by no means crucial.

Although similar in their nature, there are often differences between the ways in which cable systems of different sizes are engineered, and the two major divisions are usually considered to be into MATV and CATV systems.

MATV, master aerial television, generally refers to small distribution systems providing a service to blocks of flats and maisonettes, to office blocks, and even to small housing estates.

Although early MATV systems used to use VHF distribution, over recent years, since UHF transmissions have become the norm in the UK and as UHF equipment has improved in performance, more and more new MATV cable system builders are

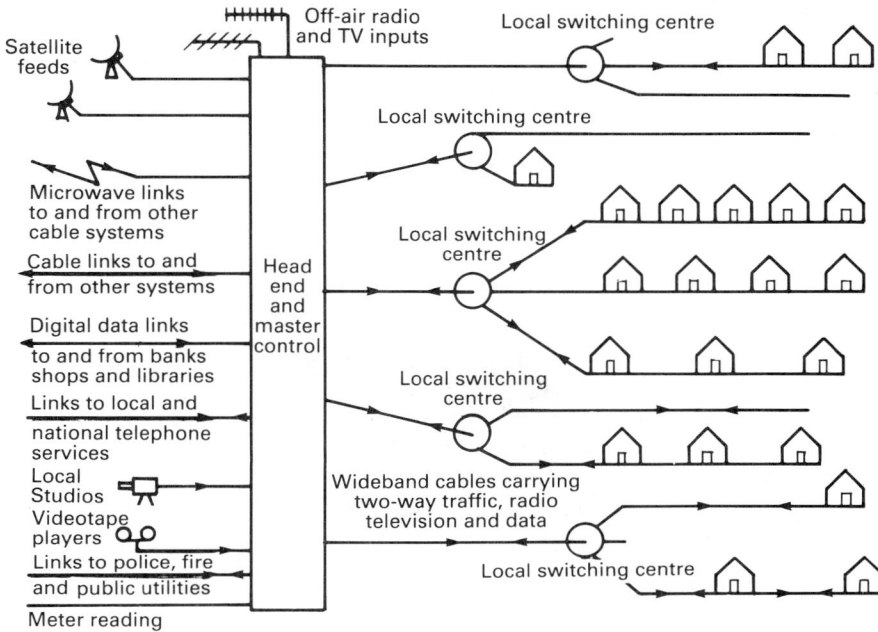

Fig. 1.2 — Wired network of the future.

choosing the simpler option of distributing television signals at UHF, without frequency conversion, for their small- and medium-sized systems.

Even with these small systems it can sometimes prove necessary to change to different frequencies within the UHF band, to avoid strong interfering signals or problems with pre-imaging. This is a form of ghosting caused by signals which reach the receiver by direct off-air pickup, perhaps through the wiring of the receiver, being slightly displaced in time from the signals on the same channel that are being sent to the receiver via the cable distribution system. This tends to occur in areas close to a transmitter where strong signals are available, and is a common problem encountered when installing MATV systems in hotels and apartment blocks in many major cities. Provided that not too many programme channels have to be converted, it is usually possible to find alternative UHF channels for the cable system which are far enough away from the off-air signals to eliminate the problems.

CATV, community antenna television, is the term applied to fairly large systems which distribute television signals to large numbers of homes in towns and villages. Such systems usually contain special compensating equipment to ensure that the picture quality is not degraded as the signals travel along the length of the system, which can cover many miles. CATV is the modern term for what used to be known colloquially as 'television relay', and when discussing 'cable television' it is usually CATV-type systems that we shall be considering.

The term 'communal television' is used in third-world countries such as India to describe a television system which uses a master antenna to feed one or more centrally located television receivers which are set up in public places so that villagers can gather together to watch the programmes. This should not be confused with CATV (Fig. 1.3).

Fig. 1.3 — Photograph of communal television system in India (*ABU Technical Review*).

Another set of inititals which the reader may sometimes encounter is CCTV, standing for closed-circuit television, which is any system where television signals are generated, distributed and received entirely within the users' own premises. Such systems are frequently used for security purposes in major office complexes and on factory sites (Fig. 1.4), and the television pictures, often of a lower quality than would be demanded of broadcast television, are sent along cables to the office where the security guards sit, enabling them to maintain constant surveillance over many different parts of the site. Cameras responsive to infra-red radiation are sometimes used to survey potential intruders without their being aware that they are being watched. Road traffic management systems often use CCTV to allow police to monitor traffic congestion on motorways and at busy road intersections, and systems

Camera observing
main entrance

Main entrance

Monitors

Video
switcher

Camera
observing
entrance

Security staff
office

Staff
entrance

Factory

Camera observing
goods entrance

Goods
entrance

Camera
observing
car park
(infra-red for
night surveillance)

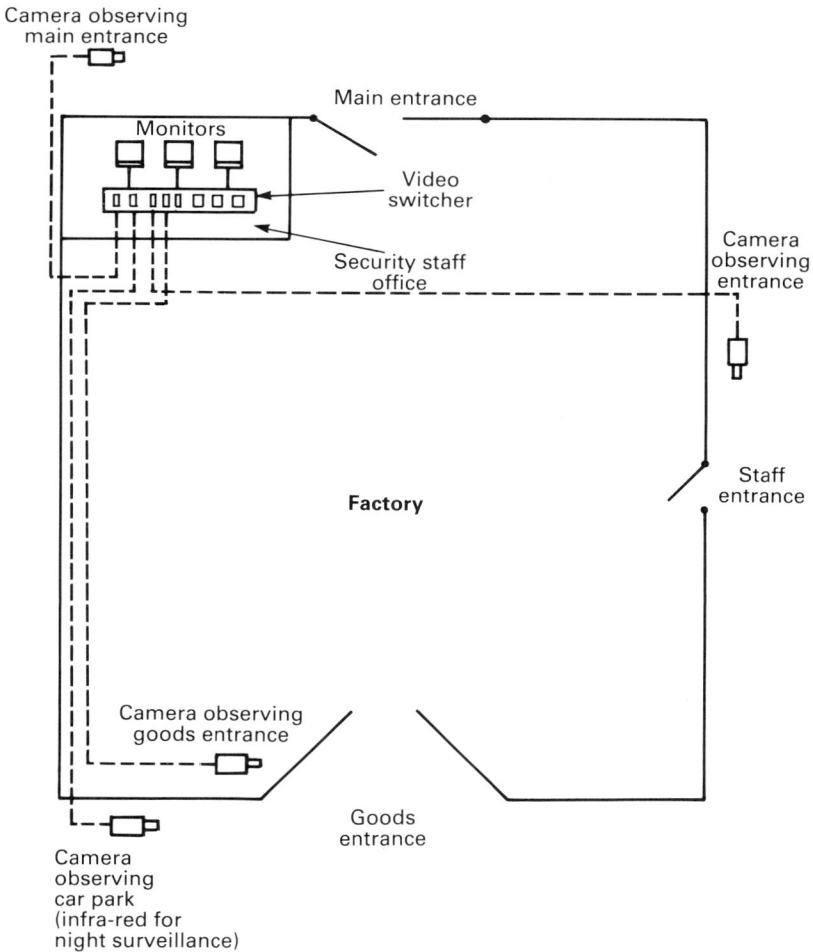

Fig. 1.4 — CCTV security system to guard factory premises.

have even been developed which can read the licence number plates of moving vehicles. Closed-circuit systems need not be thought of as unsophisticated, since many of them have been among the first cable systems to make use of fibre-optic technology. CCTV is a fairly specialist field, outside the range of this book, and we shall not consider it further.

1.1   THE RANGE OF DIFFERENT TECHNOLOGIES

The four major cable distribution technologies will be considered in some detail, starting with the old HF (high frequency) systems and then moving on to VHF (very high frequency) and UHF (ultra high frequency) systems, and to the optical transmission technologies that are likely to play a growing part in the cable systems of the future.

1.1.1 HF multipair distribution systems

Each of the television signals received at the master aerial — or head end, as the combination of aerials and their associated amplifiers is often known — is translated from the incoming UHF or VHF frequency to a frequency in the band between 3 and 30 MHz, usually a frequency of around 10 MHz being chosen. Each programme is carried on its own discrete balanced pair of wires in a multi-way cable which normally contains up to 12 pairs, and each uses the same high-frequency 'carrier' (Fig. 1.5). If four programmes are being distributed, then it will be necessary for four pairs of wires to be taken into each viewer's home, and the viewer then selects the channel which is wanted by means of a switch connected to these pairs of wires. This selector switch unit is often mounted remotely from the receiver, and it can also contain the mains switch and a volume control in some instances. The receiver in the viewer's home has to be a non-standard model which can receive the HF carrier signals. This gave rise to a lot of unhappiness in the retail trade at one time, since dealers felt that customers who were connected to one of the HF cable systems had little choice but to buy or rent the special HF receiver from the cable-operating company, thereby limiting their own chances of selling conventional receivers. Eventually it became the usual practice for the cable operator to make available an adaptor, sometimes called an invertor, to enable standard UHF television receivers to be used.

On these systems, sound signals can be translated to an appropriate HF frequency and passed through the system in the same way as the pictures, and sometimes extra sound radio programmes are carried over the same pair of wires, but it is more usual for the television sound signals to be distributed directly at audio frequencies on a balanced pair of wires.

HF distribution systems are capable of providing excellent quality sound and vision, and they are very suitable for the coverage of large areas since signal losses in cables are generally very low at the frequencies which are used, and the number of repeater amplifiers is therefore kept to a minimum. The number of channels which can be transmitted is basically restricted to the numbers of pairs of wires available in the cable, and it is not usually a practical proposition to increase the numbers of channels available once the system has been installed. This installation, together with the need for fairly complex head-end equipment with frequency convertors and filters and the unpopularity of having to use non-strandard receivers or adaptor boxes, has rendered HF systems obsolete, and no new systems of this type are being constructed.

1.1.2 VHF distribution systems

Radio frequency signals received at the head end, whether VHF and UHF, are converted to frequencies in the VHF band between about 45 and 225 MHz, and these VHF signals are fed to the distribution network which is made up from coaxial cables. Nowadays, VHF distribution systems are usually used only on large CATV systems which cover extensive areas. The main reason for converting to VHF is that VHF signals of this type can typically travel distances of up to a kilometre or so along

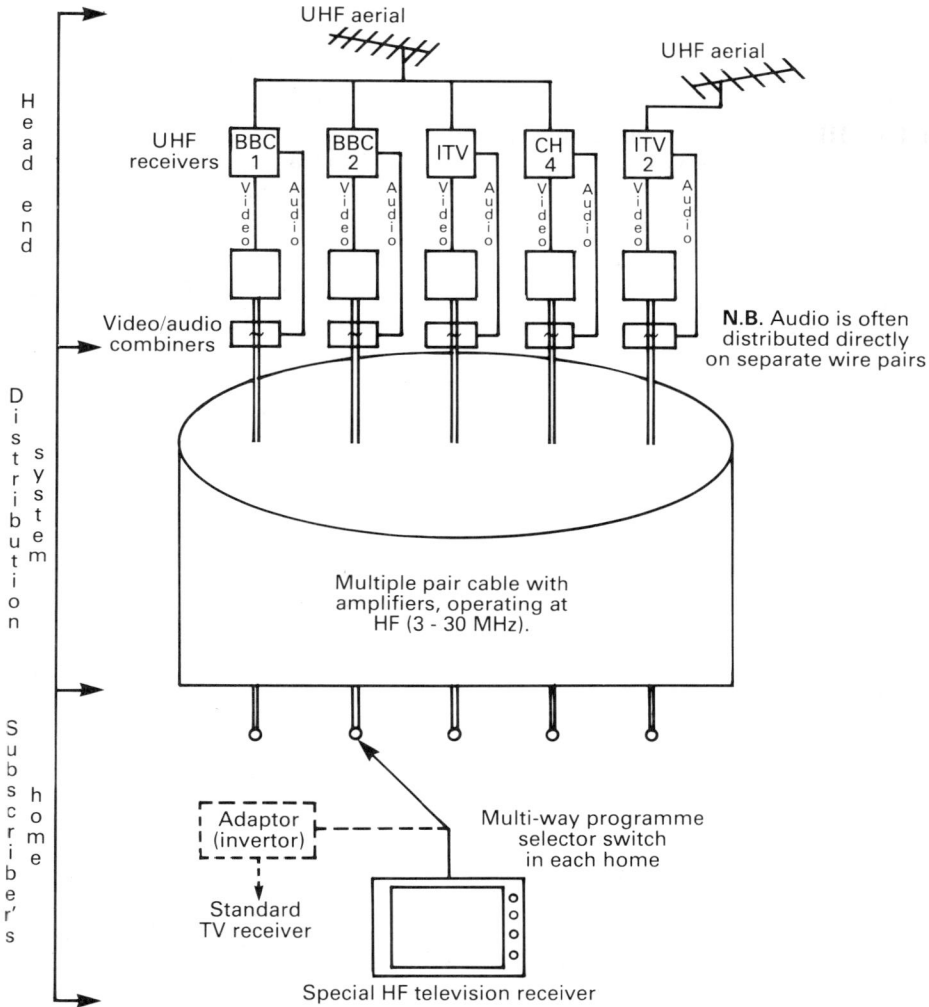

Fig. 1.5 — Typical HF cable distribution system layout.

coaxial cables without amplification, and so it is usually more economical in terms of the provision of equipment to use VHF rather than UHF, where cable losses will be much higher and amplification equipment will be required much more frequently along the long cable run.

In the UK the actual distribution frequencies used in any particular system have to be agreed with the Department of Trade and Industry so as to minimise the possibility of interference to and from other users; standards for immunity to interference and for radiation from cable systems are considered in Chapter 12. In the UK, ten channels are officially recognised for VHF distribution purposes, as shown in Table 1.1, although it is permitted to vary these frequencies slightly to avoid interference in some circumstances, and it is also possible to use special techniques to allow the distribution of more channels.

It should be noted, however, that factors other than straightforward channel usage must often be considered when trying to work out suitable frequency allocations for use on cable systems. As an example, cognisance must be taken of the facts that the normal television receivers that are in use in the UK are not fitted with VHF tuners and can therefore only operate in UHF bands IV and V; the selectivity of these receivers is generally so poor as to be totally inadequate for adjacent channel operation and operation on 'image' channels (N+9); there is so much local oscillator radiation that operation on channels five channels away from the required channel (N+5) is impracticable, since an adjacent receiver may be radiating significant quantities of local oscillator signal on a frequency of 39.5 MHz (the usual IF in the UK) above that of the required channel, which is likely to interfere with any signal at 40 MHz, i.e. five UK television channels, from the required frequency.

It will be seen from Table 1.1 that television sound transmissions are allocated carrier frequencies 6 MHz above the vision frequencies. Cable distribution services of this type normally also provide a choice of VHF sound radio programmes which are distributed on their original Band II VHF frequencies. The European over-air channel allocations for Bands I, II, III, IV and V are detailed in Table 1.2.

It can be seen that UHF television channels are 8 MHz apart in the UK, where CCIR system I is used, but different countries use different systems, which may well involve different channel spacings, so it is important to establish the parameters of the system being used before trying to design a cable network (Table 1.3). Full details can be obtained from CCIR report number 624-2 [1]. In the USA, the normal system M transmissions use 6 MHz channel spacings, and in much of Europe, where various different television systems are used, 7 MHz channel spacing is common.

Table 1.4 shows how in the USA the spectrum from 54 to 400 MHz is divided into different segments, and such an arrangement is often referred to by just its upper frequency limit, so that the system in the table might be called a 400 MHz system, or alternatively it might be called a 52-channel system. Similarly, a system with an upper frequency of 300 MHz is sometimes called a 35-channel system. The channels numbered 2 to 13 inclusive are frequently known as the standard VHF channels since these were the first to be used, the others being later additions. These standard channels also have the advantage that they can be received on any standard television receiver, without a convertor box having to be added, at extra cost to each receiver. This can be a powerful economic argument for a small and impecunious cable operator to restrict his system to 12 channels. It is also worth noting that the original 12 channels were chosen in such a way as to minimise potential interference caused by intermodulation products generated by the multiple signals. Adding the extra channels has made life far more complicated when trying to minimise the various possible forms of intermodulation interference.

A fairly simple method of providing more than 12 channels without becoming involved in all manner of frequency-extension problems is to put two cables, each capable of carrying 12 channels, around the network, instead of one. The two cables work completely independently of one another, but each carries a different set of 12 programmes on the same standard frequency channels from 2 to 13. The customer

Table 1.1 — VHF cable distribution channels in the UK

Channel code	Vision carrier frequency (MHz)	Sound carrier frequency (MHz)
A	45.75	51.75
B	53.75	59.75
* R1(B+2 MHz)	55.75	61.75
C	61.75	67.75
D	175.25	181.25
E	183.25	189.25
F	191.25	197.25
G	199.25	205.25
H	207.25	213.25
I	215.25	221.25

N.B. Channels may be offset to reduce interference, and they should then be referred to as Channel 'A+0.5 MHz'.
* Channel R1 was adopted before large offsets became common and is not usually specified for new systems.

Table 1.2 — Broadcast bands for television and VHF radio (band limits vary somewhat in different countries)

VHF Band I	Television	41–68 MHz	Not used for TV in the UK at present.
VHF Band II	FM sound radio	87.5–108 MHz	
VHF Band III	Television	174–225 MHz	Not used for TV in the UK at present.
UHF Band IV	Television	470–613 MHz	
UHF Band V	Television	615–890 MHz	

Table 1.3 — Characteristics of some TV systems used in different countries

System	Lines	Channel width (MHz)	Vision bandwidth (MHz)	Vision/sound separation (MHz)	Vestigal sideband (MHz)
B	625	7	5	+5.5	0.75
G	625	8	5	+5.5	0.75
I	625	8	5.5	+6.0	1.25
M	525	6	4.2	+4.5	0.75

Table 1.4 — VHF cable distribution spectrum in the USA

Description	Channel number/letter	Frequency range (MHz)	Number of channels
Low band	2–6	54–88	5
Mid band	14–22*, A-1	120–174	9
High band	7–13	174–216	7
Super band	23–36*, J–W	216–300	14
Super band	37–53*, AA–QQ	300–402	17
		Total	52

* Cable systems numbering of older letter channels A–W and AA–QQ. Not to be confused with the UHF channels (14–83, 470–890 MHz) used in TV broadcasting.

has a two-way switch in his home, which allows his receiver to be connected to either of the two cables, thus doubling his choice of programme material. This system is widely used in the USA, although it is rare for the maximum possible 24 programme channels to be provided, because in many areas there are a few strong off-air transmissions on VHF frequencies which can sometimes interfere with some of the cable channels. It is therefore wise to avoid using these channels for cable distribution, and in practice a maximum of about 20 programme channels is normally offered.

At the present time most TV receivers sold in the USA are 'cable-ready' and do include the mid-band frequency range, 120–174 MHz, for nine additional channels without the use of a convertor.

Other advantages of the dual-cable system are that once the simple and cheap two-way switch has been installed, the viewer can receive all the programmes on a standard receiver, whereas if an extended-frequency band is used, special convertors are required, costing the operator — and ultimately the customer — money. The dual system can also cope with those customers who do not want, or are not prepared, to pay for more than the original 12 channels. Although the second cable cannot really be considered as a back-up system for the first, since both carry different programme material, it is obvious that if problems do develop in one of the cables, a restricted service can still be maintained to customers by using the other one. This should keep viewers far happier than if they are left totally without a service, and it also gives the cable operator a breathing space to carry out repairs or essential maintenance, without having to shut down the system completely.

This dual-cable system has become so popular with cable operators, because of its lack of technical complications, that many consultants now recommend that two cables should be put into any system using undergound cable ducts right from the start of building a network. The argument for this is that the cost of the extra cable is minuscule in comparison with the costs of digging up the ground to insert larger ducts, or even of having to pull extra cables into existing ductwork, should the system need to be extended at any future date. Even though the operator of a new network may have no intention of ever increasing its capacity, history has shown a growing demand for more channels, and a little forethought can save a great deal of money in the long run.

Since the beginning of 1985 there have been no over-air television transmissions broadcast on VHF in the UK, and parts of Bands I and III have been allocated to mobile radio services. Other European countries continue to transmit television programmes on VHF Bands I and III as well as on UHF Bands IV and V, and the shared use of the VHF bands for TV broadcasting is currently being considered in the UK.

The output signals from a VHF distribution system are, of course, at VHF, and although this presented no real problems in the days when dual-standard VHF/UHF receivers were readily available in the UK, now that single-standard UHF receivers are the norm it is necessary to use an adaptor, commonly called an up-convertor or translator, to provide a satisfactory signal for the UHF television receiver. When up-convertors are used, even greater care than usual must be taken to see that signal levels available at the system outlets are kept within strictly defined limits, since the effective dynamic range of the whole system, i.e. the range of signal levels with which it can cope, is reduced. In those countries that use Systems B and G for television transmissions, including Austria, West Germany and Switzerland, receivers are fitted with VHF/UHF tuners as standard, and the need for convertors does not arise.

British Standard BS 6330:1983 6.2.3 recommends that systems using conversion to VHF to overcome the problems of cable losses over long transmission systems should be designed to reconvert the signals back to UHF at local distribution points, using channels that are not in use for local off-air signals. Such a reconversion would allow standard UHF only receivers to be used.

The system shown in Fig. 1.6 shows the basics of a typical present-day VHF wired network.

Even in large CATV systems, the demarcation lines between VHF and UHF are less rigid than was formerly the case, and although the increase in cable attenuation at the higher frequencies makes operation at frequencies above about 450 MHz significantly more expensive, some modern systems do make use of the whole spectrum of VHF and UHF, right up to about 860 MHz. A well-documented [2] system in Vienna goes as far as using twin cables each carrying VHF signals in the range 47–300 MHz for the relatively long distance trunk feeders, but converts these signals by multiplexing techniques in substations so that all the signals are available to subscribers over just one cable, which carries as many as 18 TV and 14 FM radio signals in the frequency band from 47 to 860 MHz.

The layouts of modern UHF systems and VHF systems are almost identical, apart from the use of the translation equipment at head ends and in the viewers' homes and the greater distances between repeater amplifiers when VHF is used. We shall therefore look in detail at a fairly typical modern wired system, which could equally well use VHF and/or UHF frequencies for television distribution, and the extra equipment needed for VHF systems will be discussed as we come to consider the appropriate parts of the network (Fig. 1.7).

At the present time, any such network is most likely to be built using high-quality coaxial cable as its main distribution medium; the implications of using fibre-optic techniques are considered later in this book.

Any cable distribution system can be considered as a combination of several different parts, which we shall discuss in turn.

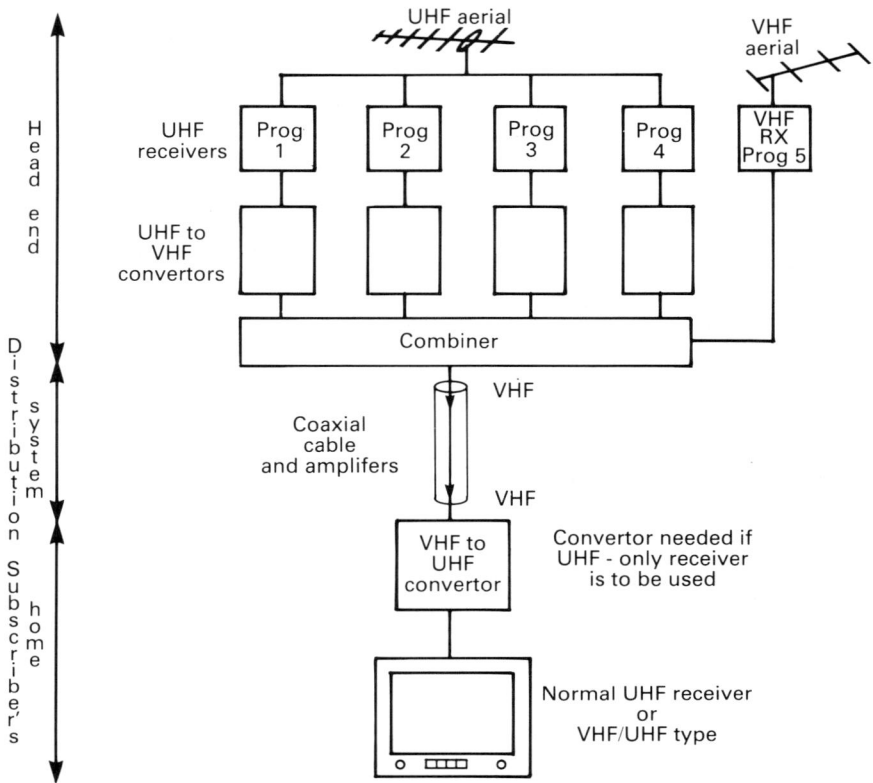

Fig. 1.6 — Simple VHF cable distribution system layout.

REFERENCES

[1] CCIR Report 624.
[2] CATV Networks in Austria and Vienna. W. Fabich. Cable Television Engineering Vol 12. No. 1. August 1982 pp. 19–21.

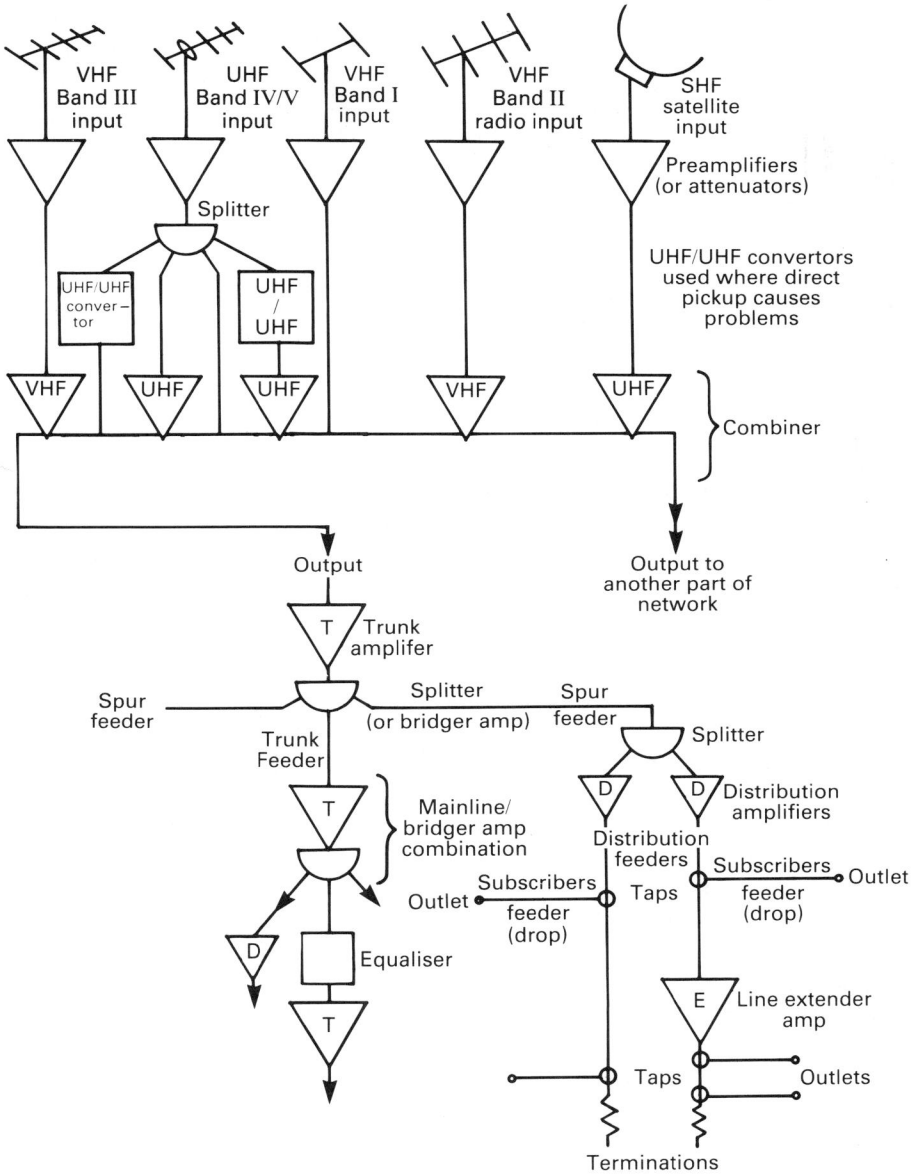

Fig. 1.7 — Typical modern wired distribution system — schematic diagram.

2

The aerial system

The aerial system will normally consist of a well-laid-out selection of aerials which have been chosen to provide optimum reception of each of the programme sources that the system will distribute. In the UK it is usual to provide one moderately high-gain UHF aerial directed towards the main transmitting station serving the locality, since this will normally provide four high-quality signals, BBC1, BBC2, ITV and Channel 4.

In many parts of the country it is also possible to receive UHF signals from adjacent regions by erecting separate aerials, and many operators use this method to provide their customers with a choice of ITV regional programmes. In earlier years there were some areas, however, where government restrictions prevented operators from picking up other ITV regions. It was felt that some of the small ITV companies might not be able to stand the financial pressures that such out-of-area reception might cause, if significant numbers of viewers chose to watch programmes from a contractor in an adjacent area, leading to a loss of advertising revenue for the smaller company. These days, competition is generally encouraged, which has led to a greater choice for viewers, although the government can still lay down restrictions in the terms of an operating licence if it believes that good reasons exist.

When the initial aerial installation is being carried out, signal strengths should be checked and the pictures should be critically examined for noise and ghosting. If satisfactory results cannot be obtained by careful positioning of a particular aerial it may be necessary to try the effect of using a different reception site. Sometimes it is found that it is not possible to receive all four UHF signals from a transmitter with equal quality, and although an improvement can usually be gained by re-positioning the aerial, in rare cases it may be necessary to install more than one aerial before all four pictures are satisfactory. It is useful to mount UHF aerials on a 'cranked-arm' so that by rotation of the crank the aerial can be moved sideways through a distance of perhaps a metre, whilst still pointing at the required transmitting station. (Fig. 2.1). This arrangement permits the optimum reception position to be easily found. The broadcasting authorities can often provide useful information about the best transmitting station to use in any particular area, and can sometimes give some idea of the

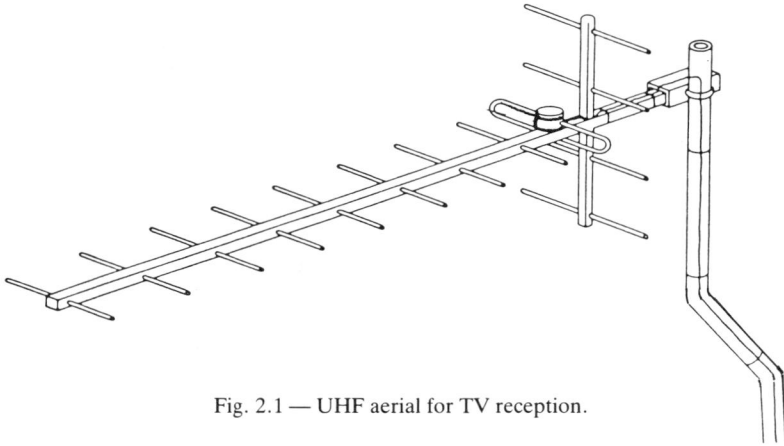

Fig. 2.1 — UHF aerial for TV reception.

practicability of receiving 'out-of-area' signals in the vicinity of the particular
receiving site.

2.1   TELETEXT

Pictures and sound are not the only things that the modern viewer demands from his
television receiver. The ORACLE and CEEFAX teletext services provided by the
broadcasters have now become an integral part of the broadcast services, and must
therefore be offered to users of cable distribution systems (Fig. 2.2). Reception of

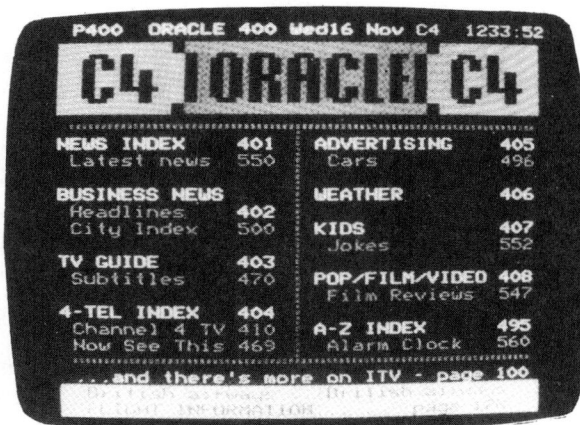

Fig. 2.2 — Teletext services are now an integral feature of cable networks.

teletext signals is often marred by the introduction of short-term, closely spaced 'ghosts' or echoes, and although pictures can be optimised by eye for minimum ghosting it is sensible to try out teletext reception by plugging the aerial outputs into a teletext receiver of known performance in order to confirm that all is well. Special teletext test equipment is available, but is expensive and unlikely to prove necessary in all but the most difficult cases of poor teletext reception. It is possible to use a normal TV oscilloscope with a delayed time-base to look at the teletext data signals, and careful examination of the so-called 'tramline' display can enable a reasonable estimate of the 'goodness' of a teletext signal to be made (Fig. 2.3). Broadcasters use

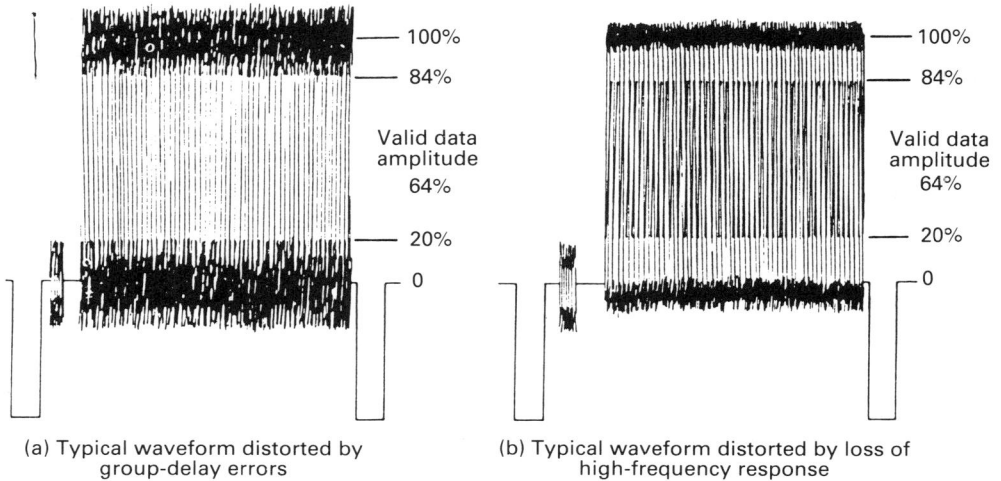

(a) Typical waveform distorted by   (b) Typical waveform distorted by loss of
        group-delay errors   high-frequency response

*Measurement of valid data amplitude by direct
inspection of Teletext waveform*

Fig. 2.3 — 'Tramline' display with oscilloscope. Photo: IBA

more sophisticated equipment to produce a different type of display that bears a vague resemblance to an eye, and the 'eye-height' of the display is taken as a figure of merit for the teletext signals (Fig. 2.4). As an example, a typical main transmitter might well be radiating teletext signals with an eye-height of up to about 70%. This will be degraded as the signal moves over its path to the receiver, and it is necessary for a receiver to be able to cope with eye-heights down to about 25%, if satisfactory teletext is to be decoded even in fairly poor reception conditions.

At present time, both UK teletext broadcasters are also providing a service called Telesoftware, which allows them to send programs suitable for home computers over the air as specially formatted teletext pages. To enable a home computer to make sense of these transmissions the signals must be received free of all errors; it would not be too disastrous if a standard teletext information page showed an occasional character error, but a computer program could well be rendered unusable if just one

Fig. 2.4 — 'Eye-height' method of teletext measurement. Photo: IBA

character was received in error. In order to overcome these problems the telesoftware pages are given extra protection in the form of a cyclic redundancy check
code, which can detect whether errors have occurred. In practice this means that the
requirements for decoding telesoftware are even more stringent than those for
decoding normal teletext signals, and it was found that a very high percentage of
schools needed to improve their sub-standard receiving aerials when the telesoftware
transmissions were first introduced, even though they had previously been receiving
television pictures that they considered satisfactory.

So far the above remarks on teletext have been concerned only with the reception
of teletext at the head end, from the aerial; we shall see later that it is also necessary
to take great care over the engineering of the complete wired system if the digital
teletext signals are to be distributed without distortion, as is necessary to provide
error-free teletext pages in the viewers' homes.

2.2 RADIO

VHF radio aerials are usually provided, and several of these may be needed if all the
BBC national radio services, and the local radio services from both BBC and
Independent Local Radio, are to be received (Fig. 2.5). In the present UK climate,
where it seems likely that new radio services will continue to proliferate, with both
community radio stations and new national radio services being planned, provision
will normally be made for the aerial tower to have the capability of carrying several
extra radio aerials.

Although it is unusual these days for medium- and long-wave radio programmes
to be carried on cable networks it is worth bearing in mind that many hours of BBC

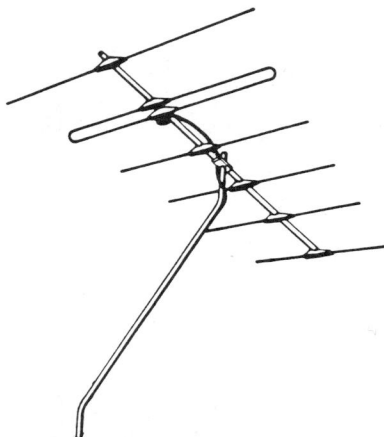

Fig. 2.5 — Band II VHF sound radio aerial.

radio are currently carried on 'split' networks, so that the Radio 4 long-wave transmitter may well be carrying different programmes from the same network's VHF transmitter. Unless the system can receive both signals, the cable listener will be at a significant disadvantage to the off-air listener, which cannot be good for the long-term reputation of cable. Medium- and long-wave aerials can consist of one or more vertical rods coupled via twin transformers and screened-twin cable to the receiver, so as to minimise the pickup of electrical noise.

2.3 SATELLITES

Medium-power distribution satellites of about 30–45 W, such as the Intelsat V series and the Eutelsat family, now provide a large number of extra programme sources for the cable operator, and although these signals can be received satisfactorily with dishes as small as 1.8 m diameter, it is generally considered wise to install much larger dishes. Many operators use dish-aerials of about 5 m to ensure that good pictures can be obtained even in the worst of weather conditions, and most agree that a 3 m dish is the minimum necessary for really 'professional' results. Such aerials must be mechanically very stable, and they are therefore usually mounted firmly on the ground, rather than on a pole or tower. Future Direct Broadcast Satellites will have powers of up to about 250 W which should allow for the use of much smaller dishes, which will require less critical mounting arrangements. In order to provide a cable service with the widest possible range of alternative programmes it will be necessary to use different satellites with different orbital positions, and this will mean the installation of several dishes pointing towards different transmitting satellites (Fig. 2.6).

It is wise to ensure that aerial installations comply with the British Standard Code of Practice for reception of sound and television broadcasting, BS 6330:1983, with BS 5640 (Aerial Specifications), and with BS CP326 (lightning protection).

Fig. 2.6 — Typical satellite reception dishes at cable head-end.

3

The head end

The head end is the usual name given to the equipment that is connected between the receiving aerials (antennas) or other signal sources and the rest of the cable distribution system to process the signals which are to be distributed.

Since all the signals which the viewer will eventually use start out from the head end, the quality of the signals at this point will determine to some extent the final quality of the received pictures, and it is therefore important to see that the head-end equipment is specified to provide first-class signals. The cost of the head-end equipment can effectively be divided between the large number of subscribers to the system, and so there is no point in trying to skimp on the quality of the equipment, and it is vital that everything possible is done to see that signal levels from the head end are optimised in order to provide the best possible pictures with the absolute minimum of noise and distortion.

Typical equipment to be found at the head end includes aerial amplifiers, frequency convertors, combiners, separators, and generators to provide the carriers upon which the various programmes will be carried around the network. Basically it is the task of the head-end equipment to take in the various signals from aerials, satellite systems, or local studios and tape machines, and to process them so that adequate signals are provided to the modulators which superimpose these signals onto separate radio-frequency carriers, one for each programme channel. All these signals then have to be combined so that they can be sent to the distribution network; this is usually done by means of a special type of amplifier with outputs that can be paralleled, or by a combination of directional couplers connected in series. Directional coupler characteristics are discussed later.

3.1 REMODULATION OF OFF-AIR SIGNALS

Three techniques for taking off-air signals and remodulating them onto the output carriers have been used in cable systems over the years. Although each of these will be briefly described, the third system, heterodyne processing, is now the most used, and is to be recommended for virtually all new large CATV systems.

3.1.1 Strip amplification
Strip amplification is sometimes called 'straight-through' amplification. As Fig. 3.1 shows, this type of system uses a set of identical broadband amplifiers which cover the whole of the broadcast band in use. No frequency conversion is involved, so the

Typical Head End Equipment Bay

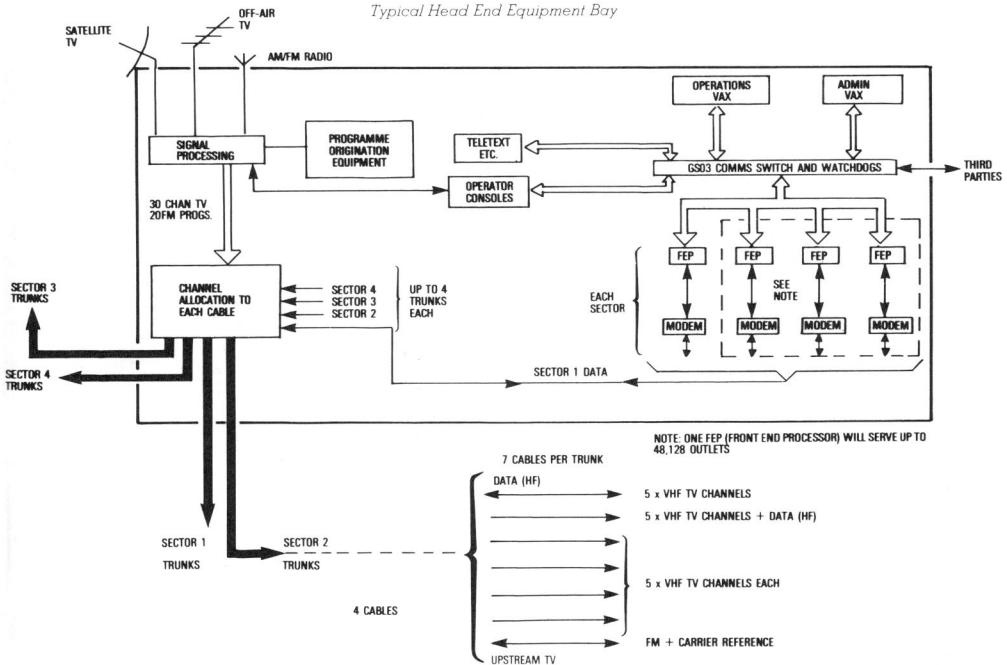

Modern head-end equipment. Photo: B.C.S.

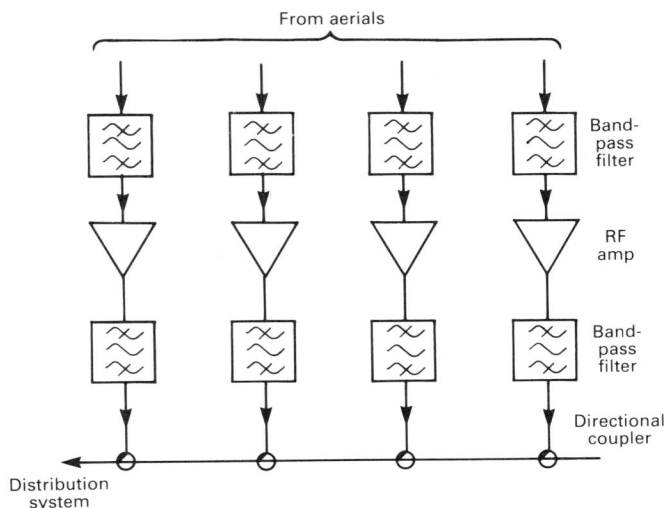

Fig. 3.1 — Strip amplifiers.

arrangement can only be used for systems where the incoming signal channel can be fed directly into the cable system. To avoid intermodulation problems, each amplifier is preceded by and followed by a band-pass filter which ideally would allow only the required channel to pass. Although such an approach can make for a reasonably economical installation, practical systems turn out to have a number of significant disadvantages. There is only limited selectivity, and since the circuitry is generally not good enough to reject all unwanted signals there are frequently problems due to interaction between the different signals, giving rise not only to cross-modulation and intermodulation distortion, but also to instability. It is difficult to apply automatic gain control, and not possible to independently control the sound signals. The strip-amplifier technique is therefore used only for moderate-quality systems where cost is of paramount importance.

3.1.2   Remodulation, or the demodulator/modulator technique
Each channel has its own receiver, more usually known as a demodulator, and the signal outputs from this are a video signal at, typically, 1 V peak to peak, and a low-impedance audio signal (Fig. 3.2). Since good-quality band-pass filtering can be used at the input to the demodulator, and the receiver itself can be designed to have adequate selectivity, the audio and video signals are usually of good quality, similar to what might be expected in a studio, and their levels can be kept constant by the use of automatic gain control, with level-limiting equipment if necessary. When using receivers with standard 'envelope' detectors, care needs to be taken to see that the group delay of the receiver matches that of a standard home receiver, since the transmitter's output will almost certainly have been pre-distorted to take account of this. More-modern designs of receiver use synchronous detectors which derive their reference frequency from the incoming signal, and these can provide better performance, which is especially noticeable when receiving teletext signals.

From aerials

RF demod   RF demod   RF demod   RF demod   RF demodulators

Video   Audio   Video   Audio   Video   Audio   Video   Audio

RF mod.   RF mod.   RF mod.   RF mod.   RF modulators

Directional coupler

Distribution system

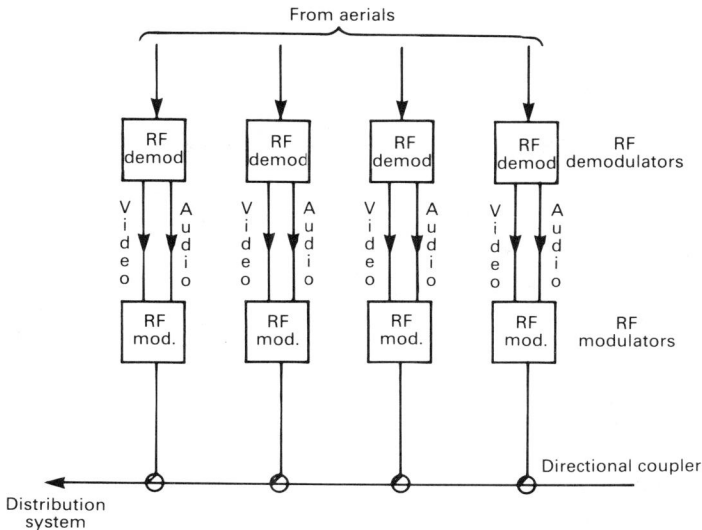

Fig. 3.2 — Remodulation head end system.

Audio and video signals can be adjusted separately if desired, and monitoring of the signal levels or of the subjective quality can be carried out at this point in the system. The audio and video signals are then applied to the input of a transmitter, the modulator, and it is the output of this modulator that is applied to the cable sytem, after being combined with other similar signals. The design of the modulator is critical if the best possible results are to be obtained, and it is usual to keep the video and audio circuitry separate until the final stage of the unit, in order to minimise any undesirable interactions. It is worth noting that video and audio inputs from tape machines or a local studio can be inserted into the cable system by applying them to similar modulators.

The remodulation technique has many advantages, including the fact that it is possible to ensure that levels of both sound and vision are kept constant, and to make any necessary adjustments to either signal without affecting the other. It is easy to make frequency changes, since the input frequency to the demodulator and the output frequency from the modulator do not need to be related, and it is possible to adjust the depth of modulation for both audio and video, which can be an advantage in circumstances where cross-modulation in the system needs to be reduced. High-quality sound and vision signals can be obtained by this technique, but since high-quality modulators inevitably require sophisticated circuitry, the equipment needed to make use of the demodulator/modulator technique tends to be expensive.

The modulators used to provide the various outputs on the appropriate VHF frequency channels need to maintain a high degree of frequency stability, to ensure that the signals do not drift out of tune, and that the designed frequency differences between the multiple channels on the system do not change, since any change could cause cross-modulation and the formation of intermodulation products and lead to complaints of unsatisfactory performance. In systems of this type it is often the case that all the frequencies used are produced by one signal generator, a comb generator, which gives out a regular spectrum of the channel frequencies at intervals corresponding to the channel separation. All the output frequencies are therefore automatically locked together in frequency and phase, and all are synchronous with the master reference oscillator in the generator. Thus only the master oscillator needs to be of the highest quality and stability, and it is relatively easy to ensure that the output channel frequencies are accurate to within a few parts per million. Permitted frequency tolerances vary somewhat from country to country, but, as an example, British Standard 6513: Part 3: 2.4.1 says that where a television signal is distributed on a frequency other than that on which it is received, the mean distributed vision carrier frequency shall be within 50 kHz of the nominal frequency for the declared channel, and that any variation from that mean shall not exceed ±12.5 kHz. Stricter limits apply when programmes are distributed using adjacent channels; in this case the total variation in frequency of each vision carrier must not exceed ±20 kHz from the nominal frequency of the channel.

3.1.3 Heterodyne processing
The most commonly used head-end processing technique in present-day high-quality CATV systems is heterodyne processing, and it will be seen from Fig. 3.3 that the

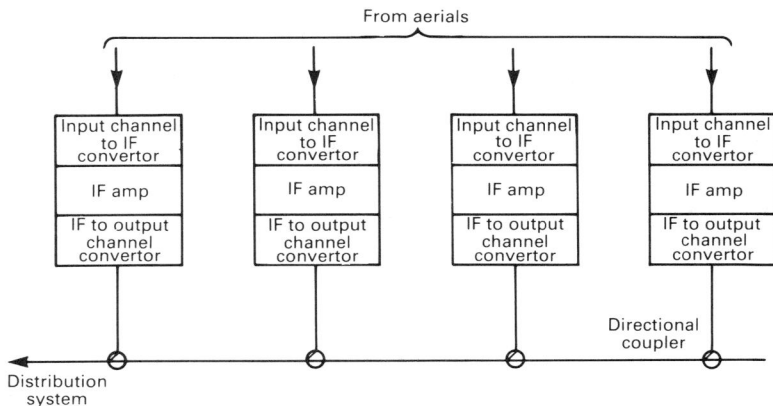

Fig. 3.3 — Heterodyne processor.

incoming off-air signals are downconverted into an intermediate frequency (IF) by beating them with the signals from a local oscillator. The IF amplification stages are designed to have excellent selectivity in order to remove any interfering signals on nearby frequencies, and to have low-phase distortion and group delay, and since all the channels will make use of identical IF amplifiers, this component can be mass-produced to a fairly high specification without costing an exorbitant amount. The IF amplifier output is then heterodyned once again with a local oscillator, so that the final output is on the desired frequency for applying to the cable. If the output frequency is to be the same as the input frequency, an arrangement sometimes known as 'on-channel operation' — the two local oscillators are locked together — is used to avoid unwanted beats which can show up as low-frequency patterning on the picture.

Such a system has many advantages, and has become the standard method of handling signals at the head end. The elimination of modulators and demodulators removes two potential sources of serious distortion. The flexibility which it allows is outstanding; if it is necessary to receive a new or different off-air signal, only the downconvertor need be changed, the IF staying constant. Similarly, it is easy to make conversions to different cable channels merely by changing the post-IF channel convertor. Programme switching, where necessary, can be done at IF, and if the operator wishes to provide emergency announcements which will appear on all cable channels, he has only to provide a single message signal on the IF. Similar arrangements can be made to allow the operator to provide a carrier, or an identification signal, even when the off-air signal disappears, either because of a transmitter fault, or after close-down of the national service. This can be useful in systems where a pilot carrier is used for automatic gain control purposes.

The IF amplifier stages can be used to control both sound and vision signals, and since no demodulation is involved, the sound-to-vision ratio should normally be the same as when it was originally transmitted. If it should prove necessary to correct the relative levels of the sound and vision carriers, this can be done in the IF stages, but it is sometimes found difficult to adjust the sound filter traps without affecting the vision, and some of the advantages of using a common design of amplifier can then be lost if too many individual adjustments are made.

3.2 AERIAL PREAMPLIFIERS

In an ideal system the signals coming from the aerials to the head-end equipment would always be of sufficient strength and with such a signal-to-noise ratio that noise-free pictures can be displayed. Unfortunately it is often necessary for cable operators who wish to provide their customers with a wide choice of programmes to make use of incoming signals from distant sources, which are less than perfect, and an aerial preamplifer can often provide a significant improvement in the signal-to-noise ratio of these pictures. We shall see later that signal levels and signal-to-noise ratios for cable systems are invariably specified at the outlet in the customer's home, which, after all, is where such things matter most.

Input levels to cable systems are not fully specified in terms of signal strength, but as an example we can note that British Standard 6330:1983.6.2.2 specifies that for a good-quality 625-line television picture, the input signal at the head end (measured in dB(mV)) should not be less than $F-8$, where F is the effective noise figure of the head-end equipment in dB(mV). In this case, if the noise figure of the head-end equipment was, say 8 dB(mV), then the input signal required at the head end would be $8-8=0$ dB(mV), which is 1 mV, a reasonable signal to expect from a medium-gain UHF aerial situated within the service area of a transmitter. In a similar manner, the standard recommends that for a VHF FM radio signal the input signal at the head end should be $F-20$ dB(mV), although this would probably be considered on the poor side by some high-fidelity fanatics.

Where these signal strengths cannot be achieved with standard aerials, it is best to try the effect of higher-gain aerials, since it is generally acknowledged that no amplifier can provide a real substitute for 'metal in the sky'. The problem with this maxim, however, is that in order to gain an extra 3 dB of signal power, or in practice a bit less, you need to double the amount of metalwork, which soon becomes impracticable when trying to receive very weak signals. On those occasions when it is just not practicable to erect even bigger receiving aerials, a suitable low-noise preamplifier fitted close to the aerial end of the feeder can provide an improvement in the signal-to-noise ratio, and therefore in picture quality. A preamplifier is therefore effective in those situations where the level of signal from the aerial is so low that the losses in the cable between the aerial and the first head-end amplifier input cause the signal at this point to have an unsatifactory signal-to-noise ratio.

Such an amplifier will only work if the preamplifer has a low noise figure and sufficient gain to override the noise of the following stages in the head end. An example of how to calculate signal-to-noise ratios is given in section 5, but at this stage of the book it is sufficient to note that a typical preamplifier might have a noise figure of less than 4 dB with a gain of from 15 to 26 dB. If too much gain is used there can be problems with self-oscillation, and since the preamplifer will invariably be mounted high up on a mast in the presence of other signals whose strength is much greater than that of the weak signal that is being aimed for, selectivity is of prime importance.

3.3   SIGNAL COMBINATION

In the final stages of the head-end equipment, all the various sound and television signals, those which are locally originated as well as those which have been received off-air, have to be brought together and combined so that they can be sent down one output cable. An ideal combining unit should ensure that each signal path is well isolated from the others, that no significant intermodulation products are produced by interaction between the large numbers of signals inevitably present, and that any losses due to the combining process are kept to a minimum. There are two common techniques in use for signal combining: paralleling, and the use of a passive combining network.

In lower-cost systems using the strip-amplifer techniques described earlier, it is possible to arrange, by careful tuning of the output circuits, that the effective output impedances of the individual output amplifiers are such that they can be connected in parallel, without any untoward mutual coupling effects, and with hardly any loss. This gives the advantage that less head-end amplification is needed than with other techniques, which can lead to lower distortion levels and a better overall signal-to-noise ratio. It is, however, sometimes very difficult to ensure that these parallel systems remain properly tuned over a long period of time, and a badly adjusted system of this type can give rise to all sorts of problems.

For these reasons it is more common these days to make use of a passive combining unit which contains a number of directional couplers connected in series in such a way as to ensure that the individual input signals are coupled to the output socket of the combining unit but are kept isolated from each other.

The basic operating principles of a directional coupler are shown in Fig. 3.4, and it is worth remembering that the prime purpose of such a device is to get signals into

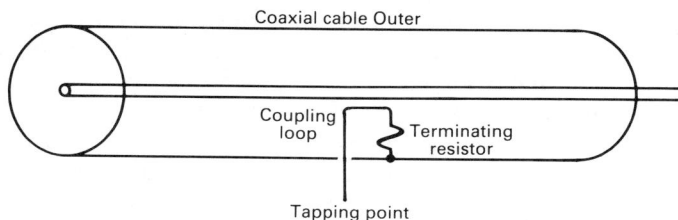

Fig. 3.4 — Directional coupler principle.

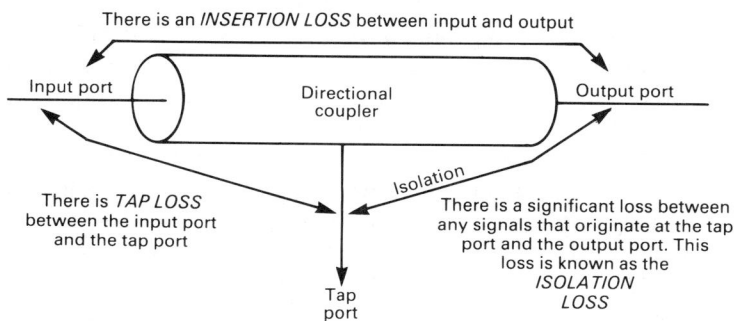

Fig. 3.4b

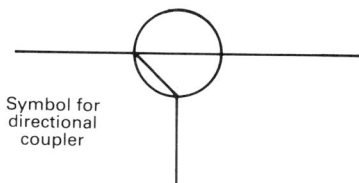

Fig. 3.4c

and out of cables with the least possible effect on the signals travelling in the main cable. Its most common use is not as a combiner, but to tap off some of the signal from the transmission line in order to feed it to the customer's outlet, and it needs to do this efficiently whilst ensuring that any spurious signals that might be generated by the customer's receiving equipment are prevented from being coupled into the main line.

Imagine the coupler as a length of standard coaxial cable into which has been inserted a small wire probe, which passes through the outer screen conductor and forms a coupling loop close to the inner conductor but not touching it. The end of the coupling loop is then connected via a terminating resistor to the outer screen of the cable.

Consider what would happen if the value of the resistor was infinitely large, i.e. an open circuit to direct current. The loop wire would act as though it were one plate of a small capacitor, the other plate being the outer conductor of the cable. This capacitor would then effectively be in parallel with the capacitive components of the coaxial line, and would charge up to a value determined by the voltage of the signal across the line.

Now take the other extreme, and consider the effect of a resistor of 0 Ω. The coupling loop can be considered to be a one-turn secondary winding of a transformer whose primary winding is the centre conductor of the coaxial cable. This time any current flowing along the line would induce a current into the probe, which would be proportional to that in the line.

Thus by varying the value of the terminating resistor, the probe can be made to respond to either the voltage across the line or the current flowing through it, and in a practical directional coupler the value of this resistor is carefully chosen so that the probe responds equally to both the voltage and the current. This means that a chosen proportion of the signal flowing along the main cable can be tapped off into the probe.

To understand how the directional coupler reacts to different signals on the main line, consider the signal travelling forward along the main line. This will induce a current and a voltage in the probe, and this current and voltage will be in phase, since there is no reason for them to be anything else. These in-phase signals will then be additive, giving an output from the probe. Any signal travelling in the opposite direction along the line, whether due to a reflection from some point of mismatch or due to a 'return-path' or 'upstream' signal being sent from a customer's house back to the head end, will lead to the current and voltage in the coupler being out of phase and they will therefore tend to cancel, giving no output from the probe. It is not possible in real life to get complete cancellation, but very high values of reduction can be achieved. It will be seen from Fig. 3.5 that because of the way in which it is

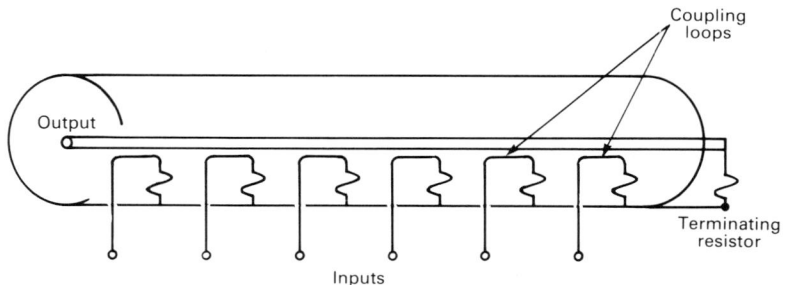

Fig. 3.5 — Combining unit using directional couplers.

constructed, such a device will ensure that the tap output and the input port on the main line are isolated from each other, which will mean that any spurious signals coming from the customer's receiver, or reflected signals due to mismatches in the cable drop to the receiver, will be very much reduced by the time they reach the main cable.

In a typical head-end combining network, shown in Fig. 3.6, a number of these

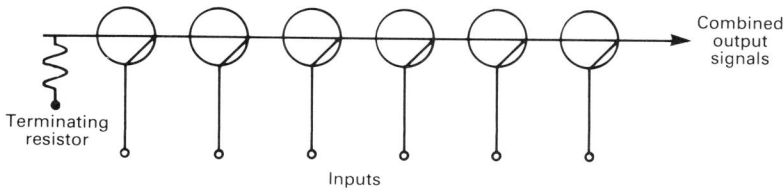

Fig. 3.6 — Schematic diagram of combining unit.

directional couplers are used, but they are installed in series, and connected in the reverse manner to when they are normally used for tapping signals off the main line. This time their directional properties are used to ensure that the input ports are coupled to the outputs, but that each input port is kept well isolated from the other inputs. Reflected signals are absorbed in the terminating resistor.

3.4 PILOT CARRIERS

In order to enable automatic gain controls to operate on amplifiers throughout a cable distribution system it is common practice to insert a so-called pilot carrier, this being an RF signal of known level, at the head end, in such a way that this signal travels along the cable system in the same way as any normal television signal. At various points around the system the level of this pilot carrier is checked automatically, and appropriate adjustments to the gain of some of the amplifiers are made in order to keep the level of this pilot signal constant. The expectation is, of course, that any variations in the level of the pilot tone will mirror the variations taking place, perhaps due to temperature changes, in the levels of the various programme signals. This is only true where the frequency of the pilot signal is reasonably close to that of the programme carrier signal, and so in systems covering a wide bandwidth it is usual to use two pilot carriers at different ends of the frequency range, often called the high and low pilots, in order to provide accurate compensation for gain changes throughout the band. The pilot carriers are frequently generated by separate oscillators which are phase-locked to the oscillator which provides the carriers for the rest of the system. The trouble with adding extra pilot carriers, however, is that these signals also act in the same way as ordinary programme signals as far as intermodulation effects are concerned, so that in a fully loaded system the presence of the carriers could cause the performance to be degraded. In order to overcome this problem, techniques have now been introduced which allow the levels of actual programme-carrying signals to act as the regulating signals for the automatic gain control of amplifiers, thus making the most of the available system bandwidth for the carrying of programme material.

4

Wired system layouts

Once all the signals have been combined they are fed to the input of the cable system proper, and it is now appropriate for us to consider various ways in which cable distribution systems may be laid out, to meet different objectives.

4.1 TREE AND BRANCH SYSTEMS

Until the fairly recent re-birth of cable technology, when cable networks first began to be seen as essential arteries for carrying the two-way streams of data that are considered to be the lifeblood of the information technology revolution, nearly all existing systems had been designed purely to distribute television and radio signals from a central point. In most of these CATV systems the signals from this central point — the head end, which was discussed in the previous chapter — are sent via a trunk feeder from which many branches are taken off in order to provide feeds to individual homes.

Fig. 4.1 shows clearly why such systems rejoice in the name of 'tree and branch' distribution systems.

Tree and branch systems are the most efficient in terms of cable and distribution equipment usage when signals have to be distributed in one direction, from head end to customer's home, which is why so many of this type are already in use. It will be seen from Fig. 4.1, though, that all programmes must generally be sent to all the homes, and this makes it difficult to arrange for individual payments to be made for particular programmes that have been received. Various methods have evolved over the years for ensuring that some programme channels are barred from those who have decided not to pay for them, but such arrangement invariably involve the installation of filters or switching equipment at each house, and this is expensive as well as being inflexible. When a customer decides to change his mind over the channels to which he wishes to subscribe, a technician has to be sent to alter the equipment in the vicinity of the house every time such a change is needed.

Another look at Fig. 4.1 will show that it would be extremely difficult and complicated to send signals back from each house to the head end and then on to

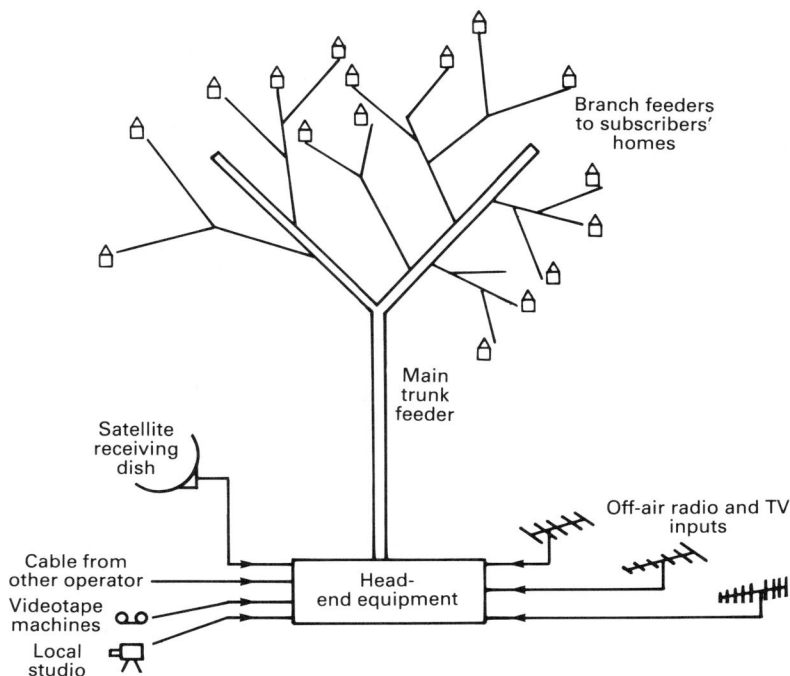

Fig. 4.1 — Diagrammatic representation of 'tree and branch' distribution system.

distant computer systems, as is required for pay-per-view television programmes and two-way data-transfer services like teleshopping the telebanking. For the satisfactory operation of these services it is essential that the customer requesting the service can be identified and charged accordingly, and that the resulting information is sent to his house and not to any other.

For these reasons then, the basic tree and branch system is unlikely to have much of a future in the brave new world of pay-television and of two-way interactive services. This does not mean that such systems cannot be sophisticated enough to provide a wide range of services, however, and Fig. 4.2 shows how one system in Vienna, Austria, manages to provide the capability for transmission of up to 18 TV channels and 14 FM radio services.

The wide range of programmes picked up at the head end is passed through two parallel-connected coaxial cables to substations, where frequency conversion takes place, converting the incoming frequencies into signals occupying the VHF and UHF bands between about 47 and 860 MHz, which standard European system B/G receivers can cope with. Each substation covers about 3000 houses, and tens of such substations can be used to serve different areas of the city.

A typical present-day tree and branch system might start off with a one-inch-diameter coaxial cable forming the trunk, and as the signal travels along this it will be attenuated, so that every half-mile or so an amplifer will need to be inserted to bring the level of the signal back up to the starting level. These amplifiers are known as

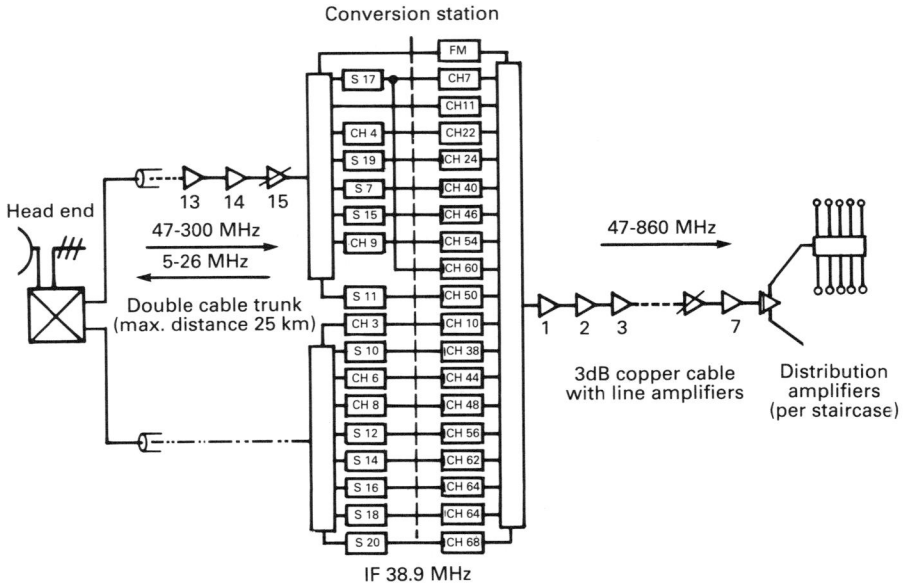

Conversion station

Head end

13   14   15
47-300 MHz
5-26 MHz

Double cable trunk
(max. distance 25 km)

47-860 MHz

3dB copper cable
with line amplifiers

Distribution
amplifiers
(per staircase)

IF 38.9 MHz

Fig. 4.2 — Basic layout of Viennese CATV network. Diagram: Telekabel.

trunk amplifiers, and since the attenuation of the cables will vary considerably with temperature it is usual for some of the trunk amplifiers to have built-in temperature-compensation circuitry, which will be designed to keep signal levels as near constant as possible. When a point is reached on the route where a number of houses are to be connected to the system, a 'bridging amplifer', often known simply as a 'bridger', is connected to the trunk feeder. Since bridgers are designed to have a high input impedance, or to be connected via a directional coupler, they impose only a small loading on the main system and they may be connected with very little effect upon the signals flowing along the trunk. The outputs of the bridging amplifer are at a high level, and are used to feed the distribution feeders that pass along the roads of the housing estates on which the customers live. Bridger outputs may be symmetrical, with equal output to each feeder, or asymmetrical, if more houses need to be served by one cable going in a particular direction than by another cable serving a different area. In some systems the signal flowing along the trunk network is divided by means of splitters to serve different areas. A splitter provides no amplification of the signal, and must introduce some loss in addition to the reduction in signal levels caused by the division into two paths. A typical two-way splitter would introduce a transmission loss of 3.5 to 4 dB at each of its outputs, so that the signal power levels out would be less than half those going in. Three- and four-way splitters are also available, but their insertion losses are even greater, perhaps as much as 7 or 8 dB.

If the distribution lines extend over more than a small distance the signals may well require further amplification before they reach the viewers for which they are intended, and distribution amplifiers known as line-extenders are then used to bring the signals back up to an appropriate level.

Once the cable is outside the customer's home, a small amount of the signal is extracted from it through either a tap or a directional coupler. It was shown in section 3.3 how the characteristics of the direction coupler make it ideal for this purpose. A low-level signal can be readily extracted from the main feeder, but the isolation provided in the reverse direction means that any signal from the receiver connection, perhaps due to a reflection caused by an impedance mismatch, will be much attenuated before reaching the main feeder, so providing valuable isolation between receiver and cable system.

Directional couplers tend to be more expensive than the simpler forms of tap, however, so most subscribers are connected via a tap. Different designs of tap are available, and different values can be obtained to ensure that the appropriate level of signal is fed to customers whatever level is available from the main cable at the point where the tapping-off occurs. Taps can be obtained to provide two, four or eight outputs, depending upon the number of homes needing to be served at each location. The connection from the tap to the home is usually made by a short length of flexible coaxial cable, commonly known as the 'drop'.

Fig. 4.3 shows the simple circuitry of some typical tap designs. The straightforward resistive tap has the best frequency-response performance but the highest insertion loss. The capacitive type has less loss because little power is dissipated in the capacitor, but can exhibit a poor frequency response. The transformer taps have good frequency response, with the lowest insertion loss and highest return loss. The addition of a matching resistor to the transformer type increases the insertion loss, which might be considered undesirable, but this effect is more than compensated for by the better return loss obtained, which means in practice that the effects of any reflections due to mismatches are minimised. This back-matched transformer tap is generally agreed to be the optimum choice of tap.

Some modern tap-unit designs use a combination of directional coupler and resistive splitters in an attempt to obtain the advantages of both types.

Once the cable drop enters the home, the signals are fed either to a convertor, which will allow the viewer to select the correct channel, descrambling it if necessary, or directly to the television receiver.

4.2 STAR SYSTEMS

In section 4.1 we saw that the nature of the tree and branch layout made it unsuitable for pay-per-view and many forms of interactive service. A perfect solution to the problem of providing for these advanced services, which really mean that the system ceases just to be a simple distribution system and becomes a complete two-way communications network, would be a star network, with the head end being at the hub of the star, and each house being fed by its own individual cable or cables (Fig. 4.4).

Such a system would easily allow the cable network operator to send each household the programmes that it wanted and was prepared to pay for, and it would be easy to charge each customer the appropriate amount for each programme or service that was provided. Two-way, truly interactive services requiring the customer to send and receive signals to and from the head end and perhaps other networks such as Prestel would be easily accommodated, since each house would have its own

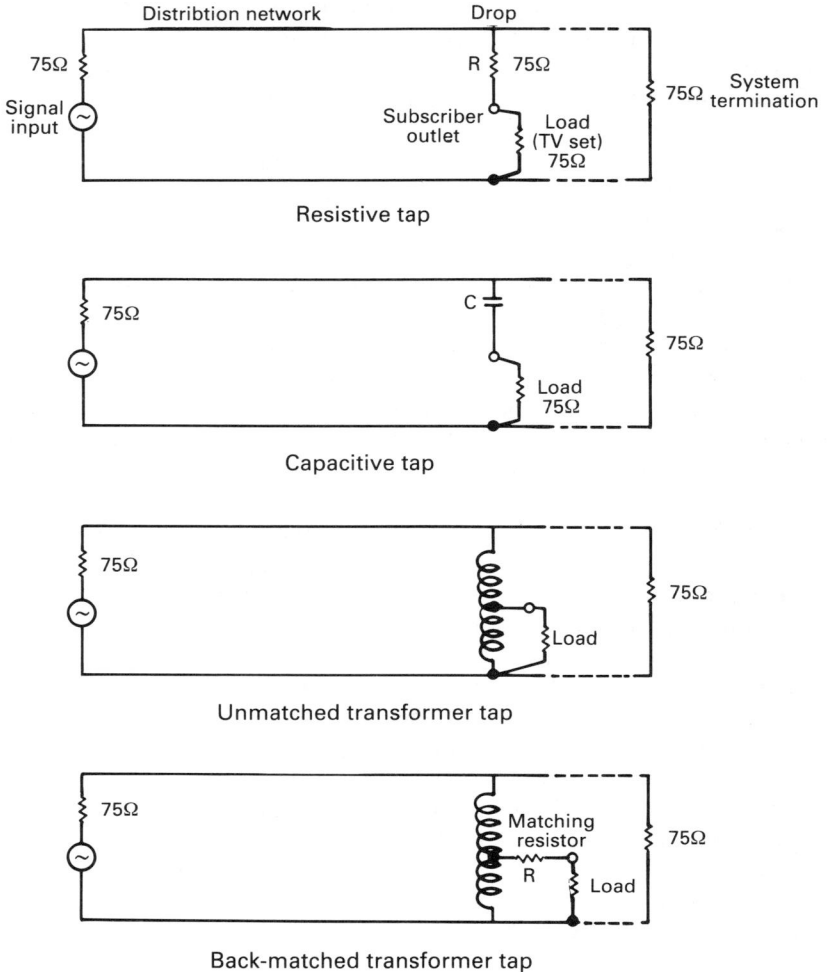

Fig. 4.3 — Different types of tap.

unique identity as far as the cable system is concerned. The one big snag with the star system is, however, its enormous cost. An individual cable or pair of cables has to be run from the head end to each house in the town, rather than many houses tapping the signals from a single trunk cable as in the tree and branch system.

4.3 SWITCHED-STAR SYSTEMS

The practical way around these problems may well be to use a combination of both the previous systems, in what has come to be called the 'switched-star' system. One system of this type that has already proved successful, and that is fairly typical of those now being introduced throughout Europe and America, was developed by that

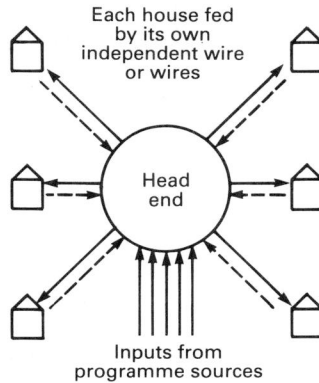

Fig. 4.4 — Diagrammatic representation of 'star system' layout.

doyen of the British cable industry, Rediffusion, whose cable interests were later to become the responsibility of a company called British Cable Services.

The Rediffusion-developed 'switched-star' system principles are illustrated below, and Fig. 4.5 shows how such a system manages to combine most of the best features of the other two systems.

In this type of switched-star system, signals are fed along a main trunk feeder which serves a number of substations, known as data concentrators. Each concentrator feeds up to eight switching centres by means of a subtrunk cable, and each of the switching centres acts as the central hub of a mini-star network serving up to 48 homes, each of which is connected to the switching centre by its own pair of cables. This arrangement allows for a reasonably efficient cable layout, whilst permitting full interactivity, and allowing the operator to charge for all the services received by a particular home. As in most systems of this type, scrambling and descrambling equipment and complex multi-channel set-top tuner/adaptors are not generally needed, since particular channels are only made available to the user when requested, the computerised switch controlling the availability of each service.

The initial design was intended to provide a choice of up to 30 television channels and 20 FM radio channels, and the single pair of cables incoming to each home allows several different programme sources to be used simultaneously in different parts of the house. The system is capable of almost unlimited expansion, and other operators using this type of system can vary the parameters according to their own needs.

As well as the usual programme origination equipment and any small continuity studio that is required, the central equipment area is equipped with two mini computers which provide and store the data which are required to keep control the switching and routing of programmes and control information. These central computers control the two-way flow of information passing through the data concentrators, but the customer's equipment receives its commands from the concentrator via the switching centre.

The main trunk feeder actually consists of seven coaxial cables bound together into a multi-coaxial cable which is about an inch in diameter. Six of the seven inner

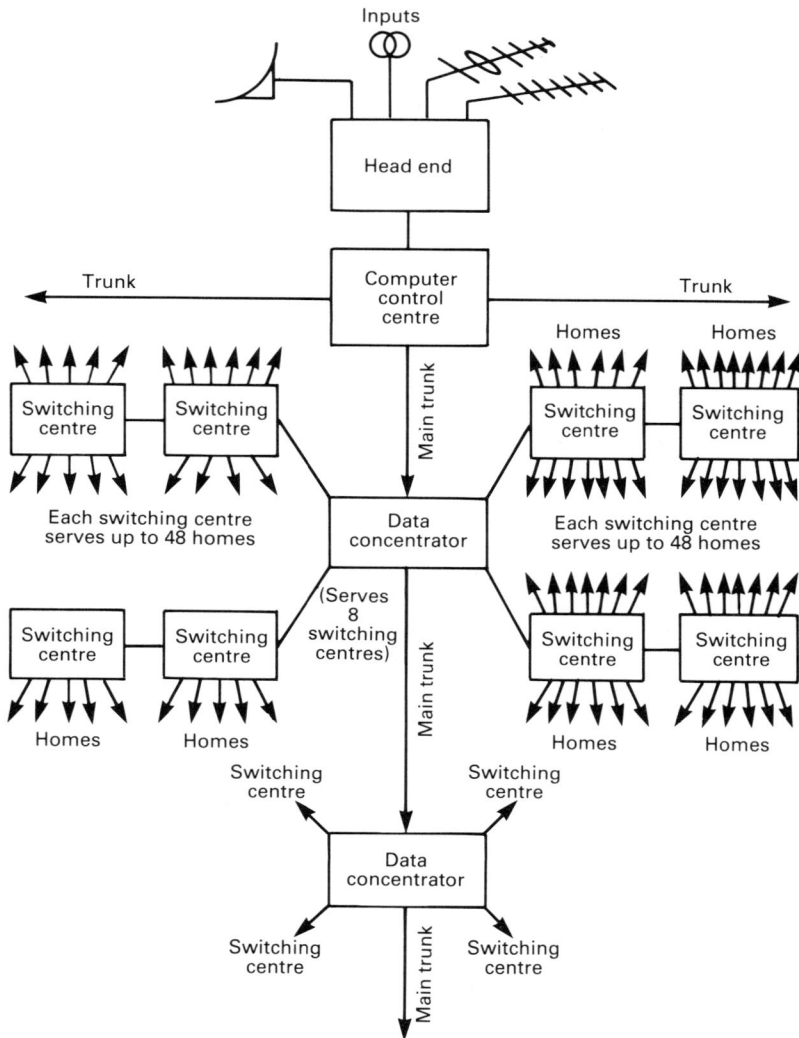

Fig. 4.5 — Simplified 'switched-star' network (based on Rediffusion system design).

cables carry up to five television programmes, each on VHF frequencies, to give a maximum of 20 channels, and the seventh is used for the VHF/FM Band II radio programmes. The VHF frequencies currently used are up to about 160 MHz, but there is no reason why higher frequencies could not be used if it proved necessary to increase the number of channels in the future.

The data concentrators are designed to cope with two-way data for up to 1152 subscriber outlet points. Each concentrator is in full communication with the central head end equipment, and acts as the main processor, a sort of telephone exchange, for data passing from head end to subscriber or vice versa. The computer store in the data concentrator has all the information about a particular customer, such as the channels that he is entitled to receive, and so it can carry out switching and billing for the various programme choices without constantly referring back to the computers in the main centre.

The subtrunk between the concentrators and the switching centres is also of the seven-tube coaxial design, carrying the signals at VHF. On receipt of the appropriate command from the subscriber, the switching centre selects the correct channel, switching being carried out at VHF by semiconductor devices. Each switching centre also contains the necessary equipment to convert the selected channels to UHF, and with 144 outlets per switch can supply up to 48 homes with three outlets in each. UHF channels in Bands IV and V are used, so the UK viewers, who generally have UHF-only receivers, can make use of the service on standard receivers without unconvertors. In countries where UHF/VHF receivers are the norm, the output from the switching centres could be either VHF or UHF, or a combination of both.

Other switched-star-type systems using similar principles are in use in different parts of the world, under names such as 'mini-star', and 'multi-star', but the actual arrangements regarding trunk layout and the usage of VHF and UHF frequencies differ in each case (Fig. 4.6). The important common feature of each is that they use a distributed network of star-type layouts with a computer-controlled switch at the centre of each mini-star. These switching centres are linked together to provide central control and monitoring, but with significant economies compared with the theoretically desirable but vastly expensive star network. The computer-controlled addressable switch is the key element in these systems, and the various manufacturers extol the virtues of their own particular designs.

4.4 SWITCHING EQUIPMENT

Even with switched-star systems it is not always economically practicable to install switches with sufficient capacity to serve every single household in a particular district, especially as it is known that it is extremely unlikely that 100% of all homes passed by the cable service will choose to pay to be connected. If, however, the switch that is installed rapidly proves unable to cope with an increase in the number of subscribers above that which was originally designed for, often called the node size, then it may well prove very expensive to install a higher-capacity switch to serve the new customers. The option of not providing the service to those customers who ask to subscribe to the system at a later date is not really open, since not only would this be extremely bad for the public-relations image of a go-ahead cable company, but in most cases it would also be in breach of the terms upon which the franchise to operate the service in a particular area was granted.

In order to overcome this problem, considerable work has been done by the statisticians of the industry to try to calculate the optimum node size for any particular network, once the overall penetration of that network is known. The penetration of a network is defined as the percentage of the population who could take the service who actually do choose to subscribe, and this figure is usually known fairly early on in the life of a new cable network. There are usually too many unknowns to calculate the optimum node size directly, but if we assume a particular node size we can, using standard statistical techniques, determine the probability of

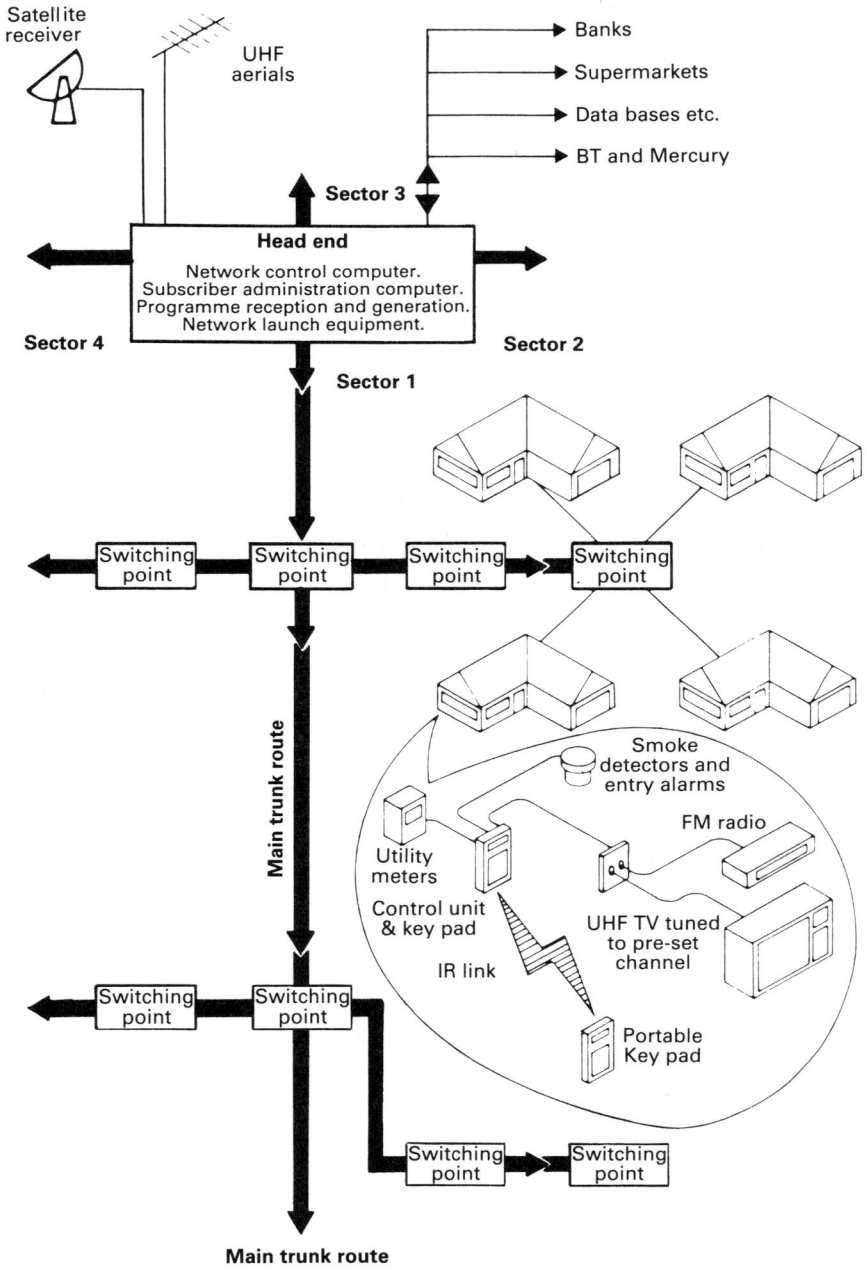

Fig. 4.6 — Layout of similar (BCS System 8) network, but showing connections inside homes.
Diagram: BCS.

Fig. 4.7 — Switching equipment. Photo: British Cable Systems.

the number of subscribers to any one switch exceeding the number for which it was designed, i.e. the node size. This type of calculation will be similar to those that have been carried out for many years by the engineers who plan the telephone network, since they have a similar problem of ensuring that the lines and the telephone exchange equipment can cope with all likely demands, whilst keeping the amount of equipment to a minimum for economic reasons. Further details of how such calculations are made are given in ref. [1].

4.5 CIVIL ENGINEERING WORKS — PLANNING

Although it is not the intention of this book to look in detail at the civil engineering aspects of cabled systems, it is worth while pointing out in this chapter on layouts that any would-be operator will need to give a good deal of thought to how his proposed system will be physically designed and planned, as well as to the electrical consider-ations discussed elsewhere. Probably the single greatest cost of any system is the cost of the civil engineering works necessary to allow a cable to be connected into each subscriber's home. A typical system will probably use ducting, either plastic or earthenware pipes, laid in front of all potential subscribers' houses, and since the costs relating to the installation of this ducting are very high, when the digging up of the road or footpath, the laying of the ducts, and the reinstatement of the road surface are taken into account, this type of work cannot be done on subscriber-by-subscriber basis. In order for the installation to make any economic sense, as well as to prevent the public outcry which would result if the cable company were for ever coming back to particular streets to dig up the road again every time a new subscriber was to be connected, the cable operator must therefore lay ducts in front of all those houses which belong to potential subscribers, whether or not they decide to take the service initially. Careful planning can reduce costs, and details such as the use of swept tees at customer tap-off points can save money as against providing street access chambers or 'manholes', as can the installation of 'pre-tubed' cables into selected lengths of road. The optimum positioning of above-ground distribution boxes or switching cabinets should also be carefully planned in advance, with due regard to likely environmental or access problems. The photographs in Fig. 4.8 show details of Windsor Cable system in the UK, and it will be seen that this installation has been carried out along the pavement rather than in the road; the contractors found that this reduced costs and allowed the cables to be inserted closer to the surface than is usually permitted under roads.

Special machines are available to carry out trenching and duct-insertion manoeuvres, but these are most suited to new developments where the positioning of the other services under the streets is accurately known, and some contractors prefer to use hand excavation in order to avoid the problems that causing damage to other services can bring. Often the public relations side of cable installation is vital to the long-term success of the scheme, since if a lot of unfavourable publicity is generated owing to street disturbance at installation time, this can give cable a bad reputation which takes a long time to live down.

Once the main cable route along the road or footpath is in place, cables are run into the houses where people wish to subscribe, either using small-diameter ducting which requires conventional narrow trenches to be dug, or by using 'thrustmoles', pneumatic impact hammers which force their way underground towards the customer's house by soil displacement. The narrow-diameter holes thus formed are ideal for the insertion of pre-tubed cables, and the advantage of this technique is that no disturbance of the subscriber's garden need take place.

4.6 FINDING BURIED CABLES

Although in a perfect world the positions of all underground services such as gas and electricity, as well as television cables, would be clearly marked on maps, this is most unlikely to be the case in practice, and one of the major problems facing a cable installer is often the need to avoid damaging existing services when trenching operations are taking place. Although little can be done about other services, it is an economic necessity for cable operators to be able to accurately locate the position of their cables at the first attempt, since it is just too expensive to excavate several holes in the road in an attempt to find a buried cable, either to find a fault or to connect another tap. Although it goes without saying that any newly laid system should be properly mapped from the outset, there are now electronic means of marking cable positions which may well prove invaluable when future operations have to be carried out by staff who played no part in the original installation.

Surface-marking techniques such as paint or stakes are relatively short lived, and so several subsurface techniques are now available. Metal detectors can be used, but they generally detect all sorts of subsurface metal and are therefore not specific enough to locate a particular cable, a disadvantage shared with other types of high-

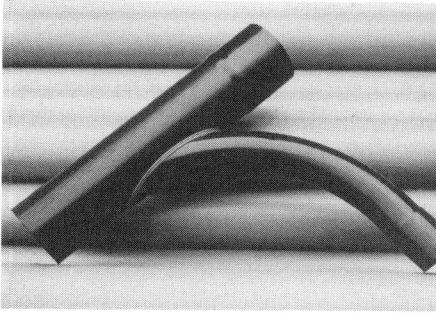

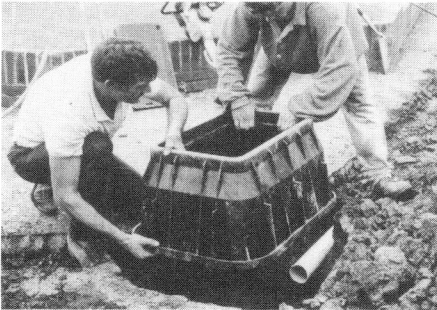

Fig. 4.8 — Photographs of civil works at Windsor. Photos: Windsor Cable, Cabletime Ltd.
Cable Television Engineering.

frequency radio locator, which work by inducing a signal into the cable sheath. One of the simplest techniques uses permanent magnets buried close to the cable, but this method has various problems, including expense, and the fact that it is not possible to distinguish between different types of buried cable. Another method sometimes used is to place 'active markers' along the cable run; the active markers are in fact low-powered radio transmitters, but these suffer from the disadvantage of having to replace batteries from time to time. Probably the most sophisticated method to date is the EMS (electronic marker system) from the 3M company. So-called 'passive direct markers' are buried at intervals along the cable run, and these are plastic packages containing an aerial system tuned to a single frequency. When an operator on the surface brings a special hand-held transmitter unit within a few feet of the cable, the marker resonates at the frequency concerned, thus providing clear identification of a particular cable (Fig. 4.9). Different frequencies can be used to

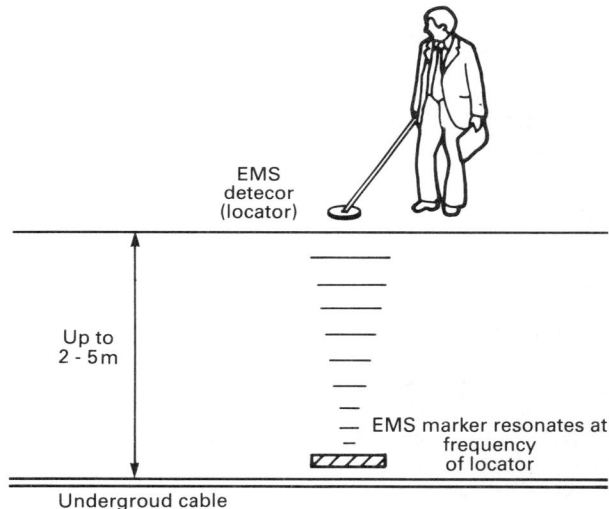

EMS
detecor
(locator)

Up to
2 - 5 m

EMS marker resonates at
frequency
of locator

Undergroud cable

Fig. 4.9 — EMS markers.

mark different cables, and colour-coded versions are available to indicate the presence of gas, electricity and telephone services. Although the use of such techniques is at present rare in CATV systems, no doubt because of the costs involved, it would seem to be a most worthwhile feature to include in new systems which are being installed, since the extra costs will be moe than recouped in the future if the position of the cable can be quickly located before digging commences.

The EMS markers can be detected at a distance of about 2.5 m from the cable, and the position of the cable is pinpointed by moving the surface unit until the field strength is highest. The manufacturer claims that such passive direct markers should have a life of more than 40 years.

REFERENCES

[1] G. R. Bandurek, Node sizes and switch sizes for star-type distribution systems, *Cable Television Engineering,* **12,** no. 10. March 1985. pp. 531–534.

5

The distribution system

We have so far discussed how the signals are received, combined onto appropriate carrier frequencies, and then switched so that customers can have the programmes that they want, or that they are prepared to pay for. Between the head-end equipment and the customer we have the distribution system proper, and we now move on to analyse the technical requirements of this part of a cable television network.

In a perfect world the signal inserted at the end of the network connected to the head end would emerge from the far end of the network in the subscriber's home with no loss of strength or of quality, but no practical system can be envisaged that comes anywhere near this ideal. Whether we use coaxial cable, the old-fashioned twisted-pair, or the latest in optical fibres to carry our signals, we find that the signals suffer from various losses and various distortions as they pass along the system, and it is therefore necessary to compensate for these factors in such a manner that the signal tapped off from the network at any given point can provide high-quality television pictures, together with sound and data of the standard laid down in the specification.

Initially we shall look at the commonest type of distribution medium, the coaxial cable, and at a later stage we will consider fibre-optic systems, but it is important to note that the same basic principles that are used in designing the coaxial system are used for the other types of network as well.

5.1 COAXIAL CABLE SYSTEMS

Since electrical signals are being passed by coaxial cables carrying the programmes from the head end to the subscriber, these signals will suffer from the normal losses that would affect any other type of electrical signal. Because of resistive losses in the cable we can expect that the signal level will be attenuated, and because of the capacitive and inductive components that effectively go to make up a transmission line of this type, we would expect some frequencies to be affected more than others, giving rise to frequency/amplitude distortion. Even though a coaxial cable has a mesh sheath designed to prevent the entry and exit of radio-frequency signals, this does not provide complete protection, and any practical length of cable is likely to be subject to electrical noise from outside sources, which will show itself as noise added to the signal being passed along the cable. We therefore need to do something about these problems, and the first step is to compensate for the resistive losses by the insertion of amplifiers at certain intervals along the length of the cable.

Fig. 5.1 shows how each amplifier is given just enough gain to overcome the

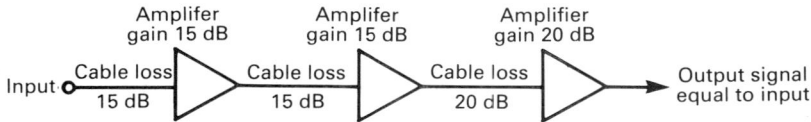

Fig. 5.1 — Part of a cable distribution system.

attenuation loss of the section of cable that precedes it, and if this could be achieved in practice it might be thought that signals could be sent along cables of infinite length without loss, an idea that we shall soon see to be fallacious. At its simplest, a cable run which gives rise to a loss of 20 dB would be followed by an amplifier of 20 dB gain, and a section with a 15 dB loss would have this loss compensated for by an amplifier with 15 dB gain. This neat idea gave rise to the now common practice of referring to lengths of coaxial cable not in feet or metres, but in units of attenuation, decibels or dB. Thus if a length of x metres of a particular cable gives a loss of 20 dB, the initial planning of the network may sometimes take no account of the actual physical length of the cable, and the planner will refer to the x-metre length as 'a 20 dB length of cable'. He will know from this that a 20 dB amplifier will be required immediately after the x metres to compensate for the signal loss. To convert the theoretical plan into a practical specification the design engineer needs to know the attentuation per metre of the cable, which he can obtain from tables supplied by the manufacturer, and he can then work out what physical distance the 20 dB cable length represents, and can site his amplifiers appropriately. To give a practical example, a typical cable might have a loss of 10 dB per 100 metres at 860 MHz, a commonly used planning frequency in the UK because it represents the highest UHF frequency which is used for over-air transmissions by the broadcasters. Thus a so-called 20 dB length of cable would have a physical length of 200 metres, after which a 20 dB amplifier would be required.

A bonus of this method of using the dB as the unit of measurement is that gains of amplifiers and cable losses can simply be added and subtracted arithmetically to obtain the total gain or loss of a system.

An idea of the problems that such a system would encounter in real life can be obtained by supposing that each of the amplifiers had a gain of perhaps 1 dB more than the 20 dB specified. After three amplifiers, the signal would be 3 dB higher than specified, providing approximately twice the power planned for, and the system could be overloaded. If each of the amplifiers had a gain 1 dB lower than specified, the signal would become progressively lower and lower than its specified level, and the cumulative effect would soon be that the signal-to-noise ratio would be lowered below that needed for high-quality pictures.

This basic principle of any transmission system, though, remains sound; we must balance the loss of each part of the cable run by a whole series of amplifiers inserted at appropriate points along the run. In addition to this, however, we must take into account the fact that each amplifier will introduce some noise and distortion in various different forms, and that the gain of an amplifier will not necessarily be exactly that which was specified. Thus we must allow for certain tolerances within which the system will work satisfactorily, and it is the calculation of the permitted tolerances and the minimisation of noise and distortion that form the most important parts of the network planner's art.

5.1.1 Noise factor and noise figure
Some meaningful measure of the practical effects of noise is obviously needed if we are to be able to carry out such design functions, and one basic term that is commonly used to indicate the noise performance, usually of an amplifier, is the 'noise factor', which specifies the degradation in signal-to-noise ratio that occurs when a signal is passed through the stage (Fig. 5.2). The noise factor is defined as the ratio of the

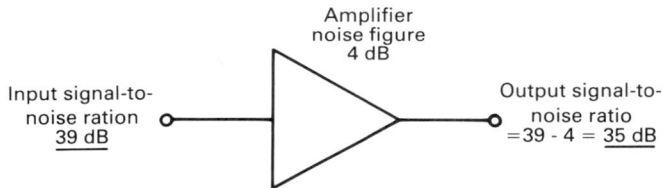

Fig. 5.2 — Relationship between signal-to-noise ratios and amplifier noise figure.

signal-to-noise ratio at the input of the stage to the signal-to-noise ratio at the output of the stage.

$$\text{Noise factor} = \frac{\text{signal-to-noise ratio at input}}{\text{signal-to-noise ratio at output}}$$

Since the units of signal power and noise power on the top and bottom of this equation are the same, the noise factor will be strictly a number, and it is important to remember this, and not to try to append some unit such as 'decibels' to the noise factor.

Because we work in decibels throughout much of the system it is often found useful to use a slightly different term, 'noise figure', which is the noise factor expressed in decibels:

$$\text{Noise figure} = 10 \log \text{S/N at input} - 10 \log \text{S/N at output}$$

where S/N indicates the signal-to-noise ratio. From this expression we can see that the noise figure is the number of decibels by which the signal-to-noise ratio is degraded when a signal passes through a stage. As an example, if the input to an amplifier had a signal-to-noise of 39 dB and the noise figure of the amplifier was 4 dB, then the signal-to-noise ratio of the output would be 35 dB (Fig. 5.2).

It is perhaps helpful in trying to distinguish between the terms 'noise factor' and 'noise figure' to think that an amplifier which produces twice as much noise power as the theoretical minimum, which we discuss later, will have a noise factor of 2, which can be equally well expressed as a noise figure of 3 dB.

It should be noted that anything that happens to reduce the signal without increasing the noise, such as the insertion of a 3 dB attenuator at the input of an amplifier, will increase the noise figure, in this case by 3 dB.

5.1.2   Equivalent input noise level

A frequently used method of considering the noise introduced by an amplifier is to consider something called the 'equivalent input noise level' of the amplifier. This sounds complicated at first, but is merely the amount of noise that would need to be present at the input of a completely noiseless amplifier in order to produce the same noise level at the output as that which is found at the output of the practical amplifier. From this definition it can be seen that the output noise of an amplifier N_{out} is equal to the input noise N_{in} plus the gain of the amplifier in decibels (A) (Fig. 5.3).

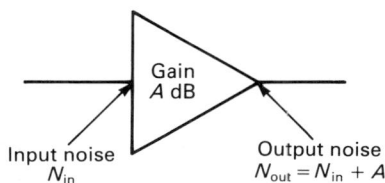

Fig. 5.3 — Output noise in terms of input noise.

$$N_{out} = N_{in} + A \tag{1}$$

It is important to remember that all electronic devices are subject to inherent thermal noise, even when they are not connected to anything, and this is usually explained as being due to the random movement of free electrons due to thermal energy. The electrons act as charge carriers, and the current which results from their rapid movement produces a voltage across the ends of the component. It is possible to gain an estimate of the level of this, the lowest possible noise level that it is theoretically possible to obtain in a system, by going back to our elementary physics books, where we find that Nyquist showed, in the late 1920s, that the average noise power P available from a conductor to a matched load is given by the expression

$$P = k\,T\,B$$

where k is the physical constant known as Boltzmann's Constant, i.e.

$$k = 1.38 \times 10^{-23} \text{ J/degree}$$

T is the temperature of the conductor in Kelvins and B is the bandwidth of the system in Hertz.

For the UK television system CCIR system I, the bandwidth over which this noise is to be measured is taken to be 5.08 MHz (BS 6513:3:1984:2.6), and for the United States system M, a bandwidth of about 4 MHz is usually assumed. Inserting these figures in the equation above, and assuming a room temperature of 68 °F, 20 °C, which is equivalent to 293 K, we obtain

$$\text{Thermal noise power for system I} = 2.05 \times 10^{-14} \text{ W}$$

and converting this to dBmV, the most usual measurement (see section 5.1.3), we obtain

$$\text{Noise in dBmV} = 10\,\log \frac{2.05 \times 10^{-14}}{1.33 \times 10^{-8}}$$

$$= -58 \text{ dBmV}$$

Similarly, the thermal noise power for system M $= -59.1$ dBmV. This figure is often called the thermal noise threshold, and if a signal were input to the amplifier at this level, the signal-to-noise ratio would be, by definition, zero because there would be zero difference between signal and noise. All practical amplifiers will introduce noise at a level greater than this theoretical threshold, and the amount of noise above the threshold is known as the noise figure of the amplifier. Fig. 5.4 indicates the different noise levels, and it can be seen that an amplifier with a 6 dB noise figure would introduce noise at a level that is 6 dB above the threshold, and that the lower the noise figure, the better the amplifier.

5.1.3 Signal level
It has been found useful to have a unit like the decibel, but which can be used to measure actual signal levels rather than ratios of levels. The impedance of television distribution circuits is invariably 75 Ω, and so a zero reference level has been chosen which is the signal level which corresponds to a voltage of 1 mV appearing across a resistance of 75 Ω. This is defined as 0 dBmV, and when this reference level is used, actual signal levels may be specified in decibels above or below 1 mV. Electrical circuit theory allows us to calculate that this reference level is equivalent to a power of

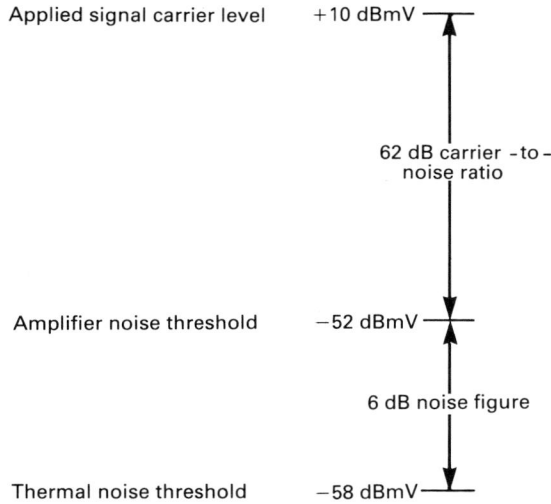

Applied signal carrier level   +10 dBmV

62 dB carrier - to -
noise ratio

Amplifier noise threshold   −52 dBmV

6 dB noise figure

Thermal noise threshold   −58 dBmV

Fig. 5.4.

$$\frac{(1 \text{ mV})^2}{75 \text{ } \Omega} = 1.33 \times 10^{-8} \text{ W}$$

In order to work out the level of a signal in terms of the number of decibels above 1 mV, dBmV, in terms of power, we calculate the number of dB the signal power P is above this reference power level:

$$N \text{ dBmV} = 10 \log \frac{P}{1.33 \times 10^{-8}} \qquad (2)$$

Returning to our equivalent input noise level equation (1)

$$N_{out} = N_{in} + A$$

we can see that the inherent thermal noise will be included in the N_{in} term, and we can in fact calculate N_{in} by adding the thermal noise to the noise figure of the amplifier in decibels.

In our distribution system shown in Fig. 5.5, the length of cable following the amplifier will introduce an attenuation equal to the gain of the amplifier A. This means that the noise due to the first amplifier at the end of this length of cable will have been reduced to $(N_{out} - A)$, which, from equation (1), is equal to N_{in}.

This signal will then be applied to the second amplifier in the chain, and this too will contribute noise equal to N_{in}.

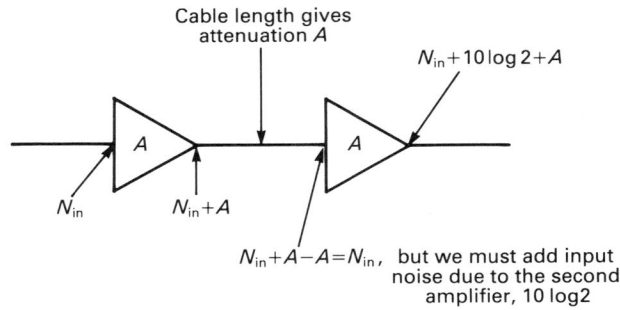

Fig. 5.5 — Noise in cascaded amplifiers.

At this point it is important to remember that these two noise powers must be added, but since we are working in dB, we cannot simply add our figures together, since the addition of logarithmic quantities is equivalent to multiplication.

In order to see how we may correctly add the two noise powers, N_{in} from the first stage and N_{in} from the second, we must go back a stage, and remember that when working in terms of power, the level in dBmV was obtained from equation (2):

$$N_{in} = 10 \log \frac{\text{Signal power } (P)}{\text{Reference power } (P_r)} = 10 \log \frac{P}{P_r}$$

The term P, in our case, will be the noise signal power, so if we are to add the noise due to the first amplifier (P) to that due to the second, identical amplifier (P), we obtain

$$\text{Total noise } N_{tot} = 10 \log \frac{2P}{P_r}$$

This is the same as writing

$$N_{tot} = 10 \log \frac{P}{P_r} + 10 \log 2$$

and therefore

$$N_{total} = N_{in} = N_{in} + 10 \log 2$$

In a similar manner we can work out that for a system with x amplifiers, the noise level at the input of the xth amplifier (N_x) will be

$$N_x = N_{in} + 10 \log x \tag{3}$$

where N_{in} is the equivalent input noise level of each of the amplifiers.

It follows from this that to obtain the noise level at the output of each of the amplifiers, it is only necessary to add the gain A of that amplifier.

Fig. 5.6 shows the practical results of the above calculations, and it can be seen

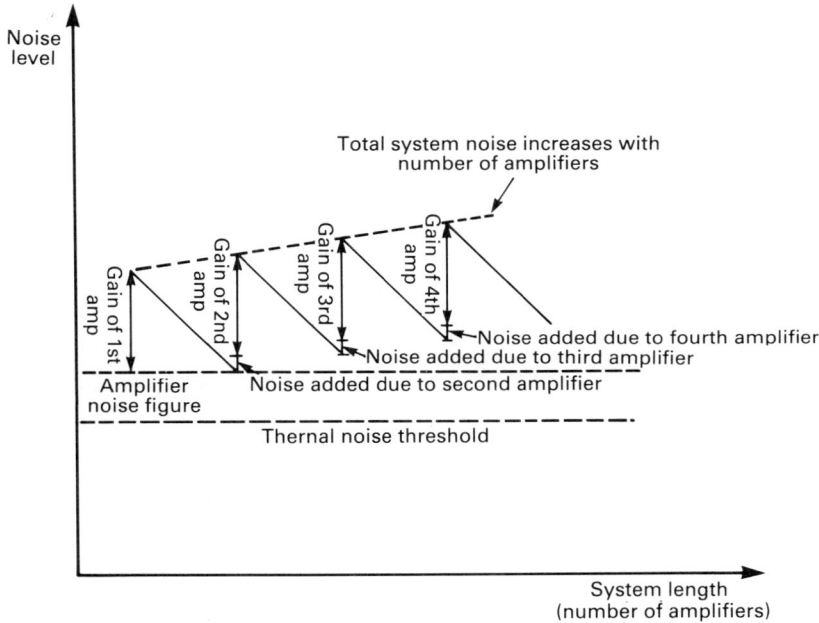

Fig. 5.6 — Diagram showing how the total system noise increases as more amplifiers are added.

that the total system noise increases as more and more amplifiers are added, which means that to maintain a particular signal-to-noise ratio, the level of the signal must be raised to cope with the increase in noise. Fig. 5.7 should make the reasons for this clear. Whatever the quality of the amplifiers, there will be a signal level which produces unacceptable distortion, and once this level is approached it is useless to go on adding further amplifiers, and thus there is a definite maximum length for any system, as was indicated at the start of this section.

Practical amplifiers are specified to provide an output signal level which ensures that the amount of distortion introduced does not exceed a particular level, but it is important to note that this will only apply where the system has just the one amplifier. It is therefore usual for an amplifier to have a variable gain control which can be turned down to reduce the gain from this maximum figure to a lower figure that will ensure that the whole chain of cascaded amplifiers works within the distortion figures that have been specified.

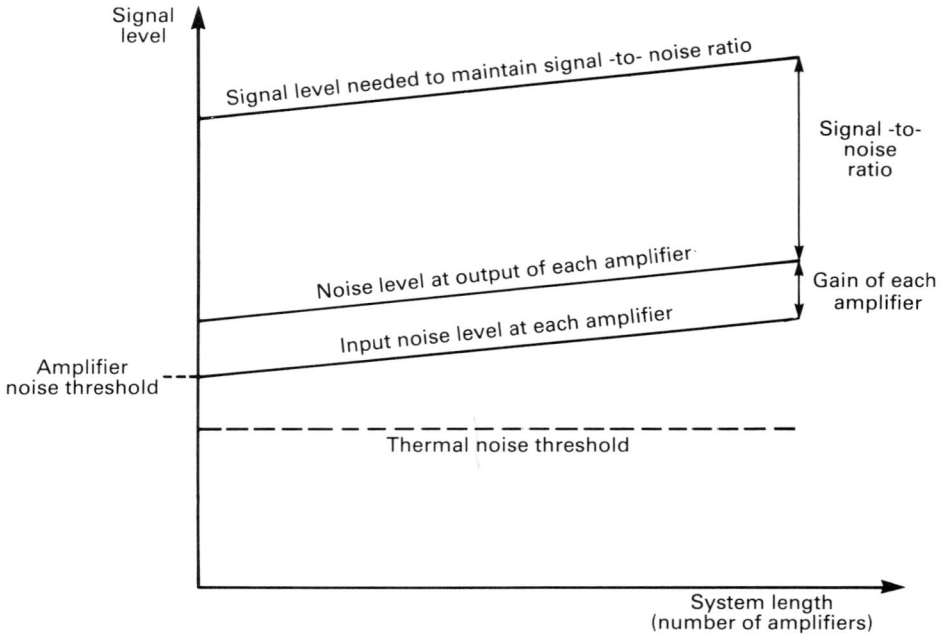

Fig. 5.7 — Diagram showing how the level of the signal must be increased as the length of the system increases, in order to keep the signal-to-noise ratio the same.

5.1.4 Carrier (signal)-to-noise ratio
In Fig. 5.7, the signal carrier level, the amplifier noise threshold, and the thermal noise threshold are indicated. When a cable system is being planned it is possible to take account of noise levels at individual points, but it is usually found to be far more convenient in practice to make use of the factor that really matters when we are considering signal quality, *the ratio between the level of the signal carrier and the level of the noise*. This signal-to-noise ratio, or more accurately in the case of AM RF signals, the carrier-to-noise ratio, is easily measured, and provides a good means of showing how the quality of a signal varies as it passes through the various parts of a system.

It is important to be clear about the difference between the use of the terms 'signal-to-noise ratio' and 'carrier-to-noise ratio', since these are often confused. When looking at cable systems as we are doing, we are concerned with the levels of the radio-frequency carrier waves that have been modulated with sound and vision signals, and hence we use the term 'carrier-to-noise ratio' (C/N). Once these radio-frequency signals have been demodulated, however, usually in the receiver, the original video and audio signals will be recovered, and these signals will inevitably have a certain amount of noise present. The ratio between the level of these demodulated signals and the level of the noise is correctly termed 'signal-to-noise ratio' (S/N). It is frequently stated that the signal-to-noise ratio of a video signal is about 4 dB lower than the carrier-to-noise ratio of the corresponding radio-frequency signal, but this can only be regarded as a useful 'rule of thumb', and the actual difference will depend on the type of signals and the modulation system used. Note that the carrier-to-noise ratio will change by 1 dB for each 1 dB change in input signal level. In section 5.1.2 we discussed thermal noise thresholds and amplifier noise thresholds, and Fig. 5.4 showed the various relationships.

From Fig. 5.8, which is a repeat of Fig. 5.4, for convenience it can be seen that the

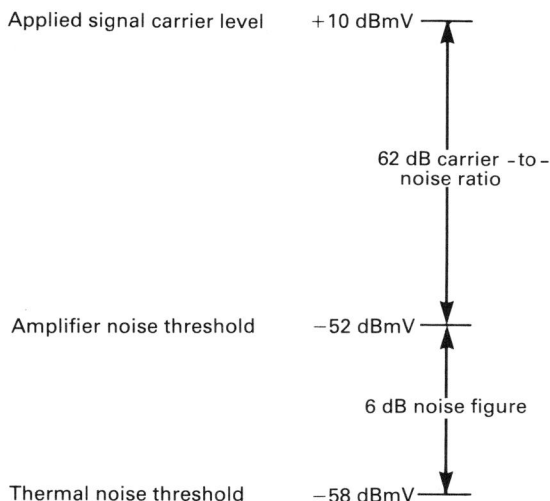

Applied signal carrier level   +10 dBmV

62 dB carrier –to –
noise ratio

Amplifier noise threshold   −52 dBmV

6 dB noise figure

Thermal noise threshold   −58 dBmV

Fig. 5.8

carrier-to-noise (C/N) for a single stage can be found from

C/N=Carrier level+58−Noise figure

Using the example in the figure, we have applied signal of 10 dBmV and a good-quality amplifier with a 6 dB noise figure. This gives

C/N=+10+58−6=62 dB

If we were to use a poorer design of amplifier, with a noise figure of 9 dB, we would need to apply a signal of +13 dBmV in order to maintain the same carrier-to-noise ratio:

C/N=+13+58−9=62 dB

We can see from equation (3) in section 5.1.3 that doubling the noise power by adding two identical amplifiers in tandem, with equal input levels, increases the noise power by 3 dB, and if the noise rises by 3 dB and the input level remains the same, this will mean that the carrier-to-noise ratio is reduced by 3 dB. Thus for every doubling of the number of amplifiers in cascade we would expect a 3 dB reduction in the carrier-to-noise ratio. This can be shown in graphical form, to make it easier to see what the effect of adding more amplifiers would be (Fig. 5.9). It should be noted that

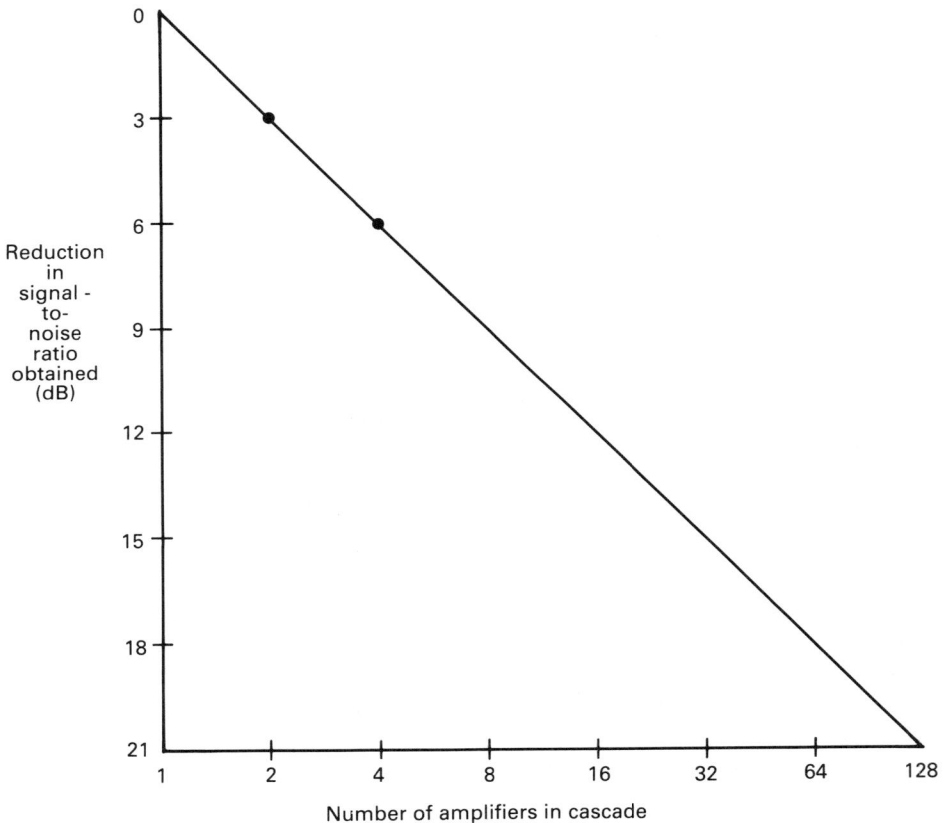

Fig. 5.9 — Diagram showing effect on signal-to-noise ratio of adding more amplifiers. The signal-to-noise ratio worsens by 3 dB for each doubling of the number of amplifiers. For example a system with 2 amplifiers will have a 3 dB worse signal-to-noise ratio than a system with only one amplifier. A system with 4 amplifiers will have a 3 dB worse signal-to-noise ratio than a system with 2 amplifiers.

if the amplifiers do not have identical noise figures, the graph does not hold true, but in practical designs, similar charts are constructed to allow the relevant figures to be read off directly.

5.1.5 Distortion

The term 'linear amplifier' is often seen in the technical specifications produced by equipment manufacturers, the term coming from the fact that for a perfect device of this type the graph of output voltage against input voltage is a straight line, since the output is directly proportional to the input (Fig. 5.10).

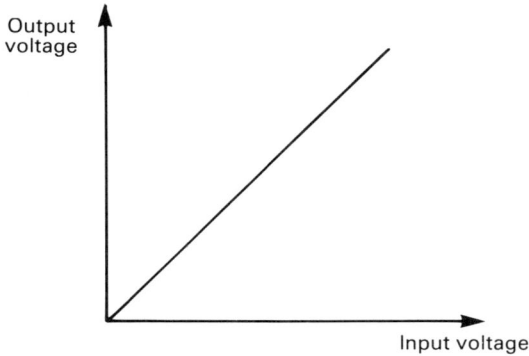

Fig. 5.10 — An ideal linear amplifier.

It is important, however, for engineers studying cable television to realise that such perfect devices do not exist in practice, and that any amplifier will only operate in a linear manner over a certain range of signal input values. Distortion will be introduced if specified signal input levels are exceeded. Fig. 5.11 shows a fairly

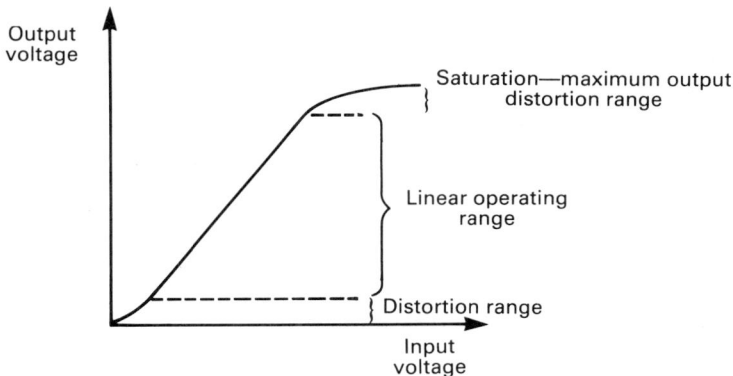

Fig. 5.11 — A practical amplifier.

typical example of the output/input voltage graph, often called the 'characteristic curve', of a practical amplifier.

Once the input signal exceeds the very lowest levels, the curve becomes linear, but after a certain input level, the rise in output voltage for a given rise in input voltage starts to fall off, and the curve bends over more and more, approaching the horizontal. From this point, any further increase in input voltage will not cause any change in the output voltage, and we say that the amplifier has reached saturation. Distortion of the input signal is therefore occurring when we leave the linear portion of the graph, and steps must be taken to see that all the amplifiers in a system are fed with signal levels that will ensure that the linear portion of the curve is used.

If the cable system were to be carrying just one signal at one particular frequency, which is never the case in practice, the output signal would therefore suffer some non-linearity distortion, which might not be too important. If, however, a non-linear amplifier or system has two or more signals applied to it, as is always the case in real life, we get a far more undesirable effect known as intermodulation distortion. The introduction of several signals to the curved part of the amplifier's characteristic curve causes them to be mixed together in a variety of different ways, which gives rise to the generation of many spurious signals. Initially these spurii generally take the form of harmonics of the various input signals, but these can add and subtract in various ways to produce large numbers of signals spaced throughout the usable bandwidth of the system. These can interfere with existing signals on the system, and lead to the superimposition of the modulation (picture or sound information) of one carrier signal onto the carrier signal of another, so that the viewer will see the pictures from an unwanted channel in the background of his wanted pictures, or suffer sound interference. This effect is known as cross-modulation.

Considering the simplest case, when two signals f_1 and f_2 are mixed together by passing them through a non-linear system, the output will contain, among other things, the original signals f_1 and f_2, their harmonics, and also signals which have frequencies equal to the sum and difference of the original frequencies, f_2+f_1 and f_2-f_1 (Fig. 5.12).

Anything that is produced in addition to the original signals f_1 and f_2 is distortion. This form of intermodulation distortion is called second-order distortion because in a mathematical treatment of the mixing of signals these spurii are related to the second term of a long equation, which is proportional to the second power, or square of the input voltage. Although there is no need for us to concern ourselves here with the intricacies of the mathematics, it is probably worth noting that the output signal V_{out}, from a non-linear amplifier being fed with an input signal V_{in}, can be obtained from an equation of the form

$$V_{out} = A\,V_{in} + B\,V_{in}^2 + C\,V_{in}^3 + D\,V_{in}^4 + \ldots$$

where A, B, C, D are constants. The first term is the undistorted output of an ideal amplifier, so the original signals f_1 and f_2 have come through even the non-linear amplifier unscathed. The effects of the second-order spurii may or may not be significant in a real system, depending upon whether they fall on top of other signals

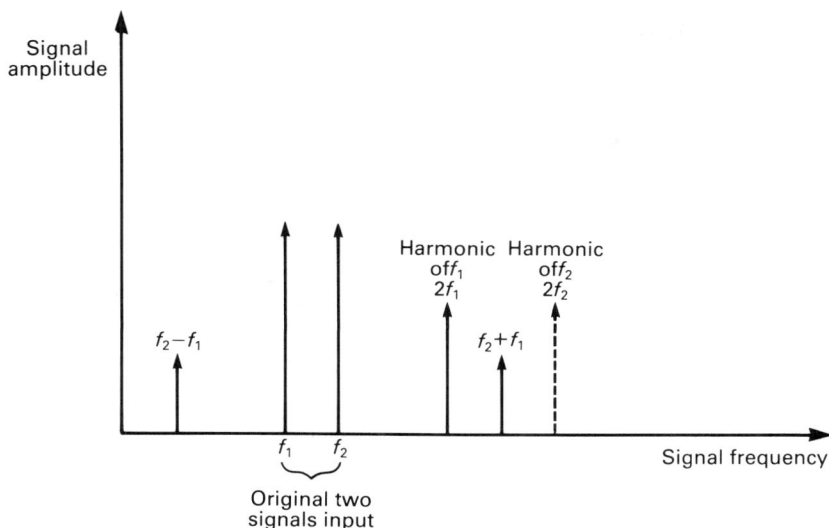

Fig. 5.12 — Showing in much simplified form some of the spurious signals generated when just two signals are passed through a non-linear amplifier.

in the system. In some small systems carrying only a few channels it is possible to arrange for the spurii to fall in channels that are not used, so that the viewer is not aware of their existence. Problems can then arise when the operator decides that he wants to add extra programme sources, because once a system is carrying a large number of programmes it is almost certain that the spurii will fall within occupied channels, causing interference. A special case of intermodulation interference is known as composite-beat interference, where the spurious signal resulting from intermodulation consists of a group or groups of closely spaced products. It is worth noting that intermodulation can be seen on a system which is unmodulated, as a beat pattern, usually in the form of diagonal bars. Significant improvements in system performance can be obtained by careful arrangement of the carrier frequencies and their spacings. It is sometimes possible to arrange for potentially annoying beat signals to fall in parts of the spectrum where they cause least visible or audible inerference, and a skilled planning engineer will be able to achieve very much better results over a complex system than someone who tries to add new frequencies on an *ad hoc* basis to an existing system.

   A far more serious type of distortion occurs, however, when the third term on the right-hand side of our equation exceeds a particular level. This third term, which is known as third-order distortion because it relates to the cube of the input signal, will cause even more spurious signals to be generated at different points throughout the frequency band being used, but this is far from being the most serious effect. Whereas with second-order distortion the original signals f_1 and f_2 were unaffected, it can be shown mathematically that the third-order terms cause the original signals f_1 and f_2 to be affected in such a way that each of the original signals is made to carry the modulation of the other signal in addition to its own. This is the cause of the cross-modulation referred to earlier, and as well as the effect of seeing one picture in the background of another, the sync pulses from other channels can cross-modulate so that vertical lines can sometimes be seen, moving across the screen, giving the so-called 'windscreen-wiper' effect. The wanted signals can also suffer from the effects of sound on vision and vision buzz on sound.

The more signals that are carried by a cable network, the more spurious signals are generated, and in a large system with 20–30 channels, many thousands of spurii can be generated, and it is obviously important to keep the levels of these signals to a minimum. It is now possible to use computers to work out the numbers and the levels of the spurious signals that may be generated in a system, in order to permit the system planner to minimise any undesirable effects before the installation of the equipment begins. Computer programs make it possible to take into account higher-order intermodulation products, which can start to be significant in systems which use amplifiers which have significant gain over a wide bandwidth.

The amount of cross-modulation in a system can be defined as the ratio between the level of the interfering signal, that is the unwanted signal, and that of the wanted signal. It is usual to take the level of the wanted signal as that which occurs when the signal is 100% modulated, and although cross-modulation is sometimes referred to in percentage terms it is more usual to express it in decibels and call it the cross-modulation ratio, which will have a negative number.

It is an unfortunate fact of life that the total distortion produced by an amplifier increases as the signal level increases, and so if we are to keep distortion to within agreed limits, we must keep the signal level below a specified maximum level. An amplifier manufacturer therefore needs to specify not only the distortion produced in this amplifier, but also the output level at which that distortion is produced. Additionally, since the number of intermodulation products will increase with the number of channels used, as we saw earlier, the specification will also need to state the number of radio-frequency signals (carriers) to which this distortion figure applies, this term sometimes being known as the 'channel loading'.

Thus a manufacturer might specify an amplifier as having an output of 55 dBmV with −60 dB cross-modulation when used in a 12-channel system. For a 35-channel system the output level might typically need to be reduced to 59 dBmV in order to keep the cross-modulation figure at −60 dB.

The cross-modulation of an amplifier at any particular output level (X_{mod}) can be calculated from

$$X_{mod} = X_{spec} + 2(OP - OP_{spec})$$

where

X_{spec} is the cross modulation in dB specified by the maker

OP is the actual output level in dBmV

OP_{spec} is the maker's specified output level in dBmV

If we insert the relevant figures for the amplifier specified above, and adjust its output level to 50 dBmV, a reduction of 5 dBmV, we obtain

$$\%x_{mod} = -60 + 2\,(50-55)$$
$$= -70 \text{ dB}$$

This is a reduction (an improvement) of 10 dB, and by similar calculations it can be shown that a reduction of 1 dB in output level causes a 2 dB increase, i.e. improvement, in cross-modulation, and vice versa. We thus find that amplifiers are subject to two contradictory requirements: if we wish to increase the carrier-to-noise ratio by 1 dB we can increase the input and output levels by 1 dB, but by increasing the levels in this way, we also make the cross-modulation distortion worse by 2 dB!

Distortion in cascaded amplifiers

In any real system we will have numbers of amplifiers in cascade, and the various distortion products will add. The second-order products add on the basis of their rms power, but higher-order products are not correlated in the same way, and it can be shown [ref. 1] that the total cross-modulation X_{tot} of a system of n amplifiers in cascade can be obtained from the expression

$$X_{tot} = X_{mod} + 20 \log n$$

where X_{mod} is the cross-modulation per amplifier in dB.

Putting some figures into the equation above for a chain of five amplifiers with an individual cross-modulation ratio of -80 dB, we obtain

$$X_{tot} = -80 + 20 \log 5$$
$$= -80 + 20(0.7)$$
$$= -80 + 14$$
$$= -66 \text{ dB}$$

If we now double the number of amplifiers in the cascade to 10, we obtain

$$X_{tot} = -80 + 20 \log 10$$
$$= -60 \text{ dB}$$

Thus we can see that doubling the number of identical amplifiers causes the combined cross-modulation to be degraded by 6 dB (in voltage terms). This can be read off from a chart of the form shown in Fig. 5.13, and in the real-life case where amplifiers have unequal cross-modulation specifications it is common for manufacturers to produce similar charts to enable the total figure to be readily determined.

Some manufacturers use another method of looking at the same thing, and

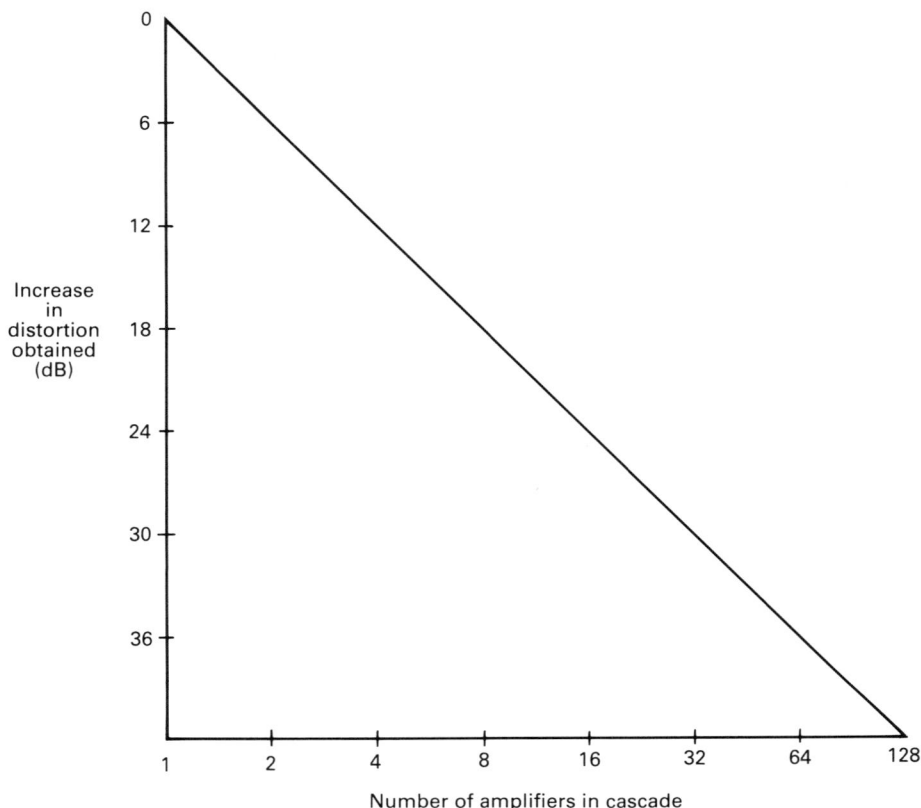

Fig. 5.13 — Diagram to show the effect on cross-modulation distortion of adding more amplifiers. The distortion increases by 6 dB for each doubling of the number of amplifiers.

consider the signal-to-cross-modulation ratio instead of the reciprocal quantity which we have used so far. This is a positive quantity measured in dB, and is usually considered in power terms. As an example, the Wolsey Electronics planning document [ref. 2] states that for a given signal-to-cross-modulation ratio, each amplifier output must be reduced by 3 dB (power) every time the total number of cascaded amplifiers is doubled. This is effectively the same rule that we used in the previous paragraph, but expressed in a different way, and the manufacturer offers 'derating charts' to help the system planner cope with combinations of dissimilar amplifiers.

5.2 PRACTICAL LIMITATIONS

We have moved a long way from our earlier ideal situation where we proposed to merely add amplifiers at certain points along the cable run which would overcome the losses of the previous section of cable, and the difficulties should by now be clear. The minimum signal level that we can use is determined by the carrier-to-noise ratio that we need, and the maximum signal that can be used is limited by the amount of distortion that can be tolerated.

This can be expressed in another way. If a system has a large number of amplifiers, the signal-to-noise ratio will decrease as the signal passes through the system, until it becomes unacceptable. Thus the number of amplifiers which can be used is limited, and hence the length of the system is limited. If we try to increase the signal levels to improve the signal-to-noise ratio then the level of distortion will rise, and will eventually become excessive. This too will determine the maximum length of a system. Fig. 5.14 shows this effect.

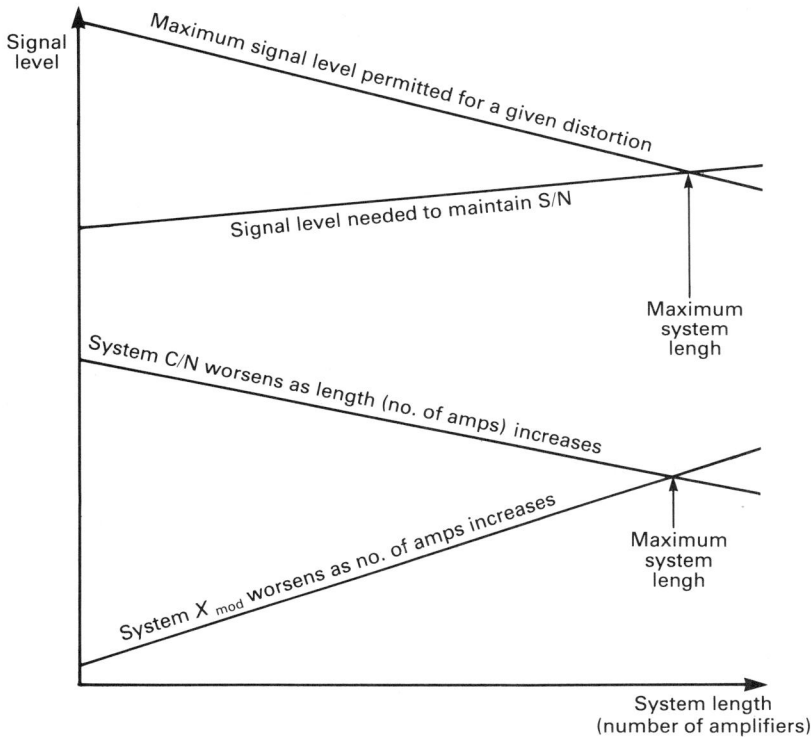

Fig. 5.14 — Factors determining maximum system length.

Maximum system length

The difference between the upper limit of signal level, where distortion occurs, and the lower limit, where noise causes problems, is known as the dynamic range of a system, and it is a useful 'figure of merit' or measure of the 'goodness' of a system (Fig. 5.15). The term 'overload-to-noise ratio' is sometimes used. The dynamic range of a system will reduce as the length increases.

From all of the foregoing remarks it will be seen that the achievable system length and the optimum spacing of amplifiers will depend not only upon the gain of the amplifier which can overcome the losses in a given length of cable, but also on the

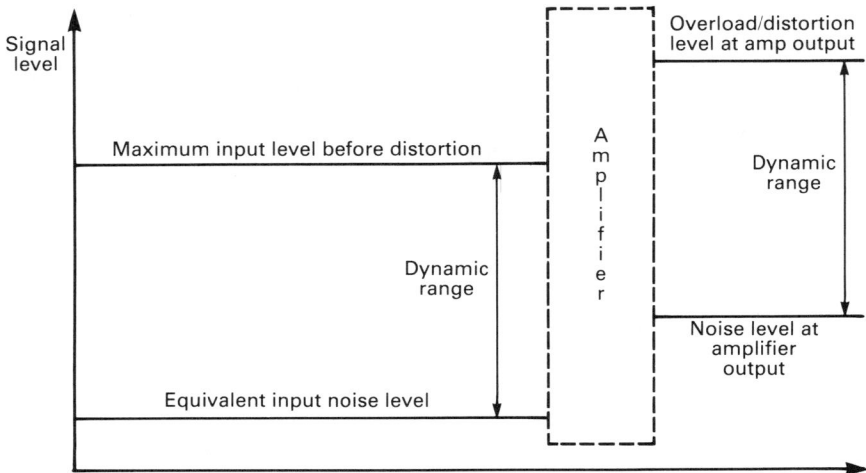

Fig. 5.15 — Dynamic range.

performance of each amplifier in the chain, and on the number of amplifiers which are to be used. At first sight it may appear that there are too many variables to enable any sensible decisions to be reached, but in practice, amplifiers are available with gains of around 15–20 dB, and once the type of amplifier has been decided upon, the rest of the figures can be calculated relatively simply. If the figures then turn out to be unsuitable, showing perhaps that the desired system length cannot be achieved, then an amplifier with more suitable characteristics can be substituted, and the figures recalculated. For any chosen type of amplifier it is possible to work out the optimum spacing, i.e. the setting of the gain control, mathematically, by considering the mathematical conditions under which the dynamic range is a maximum. This gives theoretical figures of around 8.7 dB for a single-stage amplifier, 11 dB for two stages, and 13 dB for three stages, but in practice these figures are rarely used. For a start, the gain of the amplifier will often need to be different at each of the frequencies used, and it is often difficult to compensate for signal variations due to temperature changes in the equipment. It may also be significantly cheaper to use a few higher-gain amplifiers rather than many of the optimum gain; the final system length will then be restricted, but this may well be acceptable.

5.2.1 Amplitude/frequency response

In our look at the factors affecting the basic design of the coaxial-cable-based system we have so far considered noise, and distortion, but one other factor plays an important part, and that is the amplitude/frequency response (Fig. 5.16). The attenuation of a coaxial cable is greater at higher frequencies than at lower frequencies, and it is generally reckoned that the signal loss increases as the square root of the signal frequency. Thus the loss at the lower frequency channels will be much less than that at the higher frequencies, and this means that it is not possible to

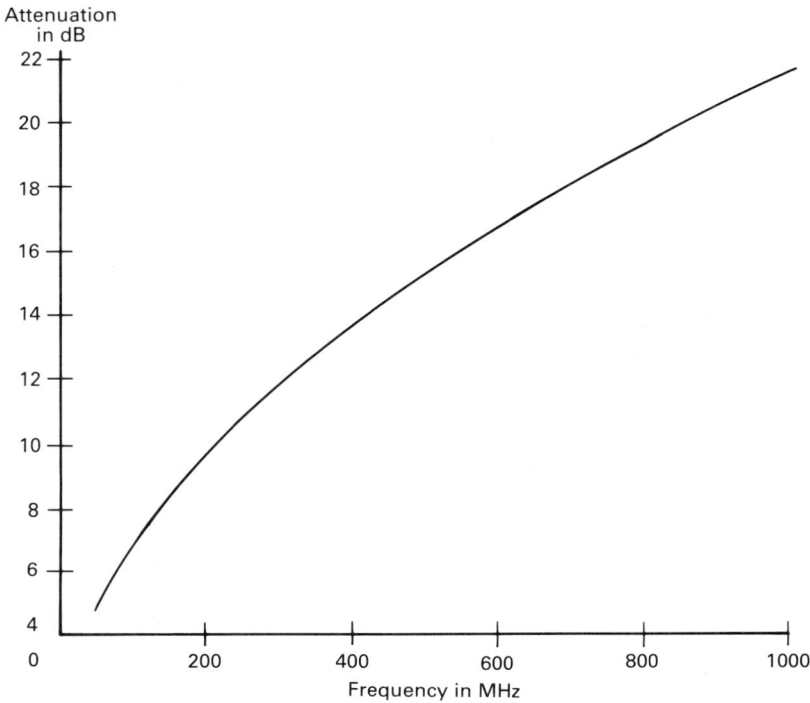

Fig. 5.16 — Coaxial cable attenuation versus frequency.

correct for the loss by simply introducing an amplifier of a certain gain, as we have been considering so far. This is because if the necessary gain is calculated for channel 21, considerably more gain will be needed to bring the signals up to the necessary level at channel 68 (Fig. 5.17). Ideally we need to give the customer equal amounts of signal on the channels which are fed to his home, and we must therefore do something to correct for the differences in the signal levels of the different channels.

Fig. 5.18 shows this effect clearly. If the input signal to the cable (i) is the same at all frequencies, the signal at the end of a length of cable (ii) will decrease with frequency. In order to remedy this we must arrange for the amplifier to correct for the cable losses at each particular frequency, a process that we call equalisation. In an ideal world the gain of the amplifier would match exactly the loss of the preceding cable at each frequency, and although we cannot achieve this degree of perfection in real life we can arrange for the gain of the amplifier to rise with frequency, as shown in (iii), so that the signal level at the output of the amplifier (iv) is once again equal at all frequencies in the band. In practice, the accuracy with which equalisation can be carried out limits the maximum number of amplifiers in a system, and hence its total length.

There are three important terms used in connection with this aspect of cable television. The variation of signal amplitude with frequency over part of the system is known as the *tilt*, the corresponding variation of amplifier gain with frequency is

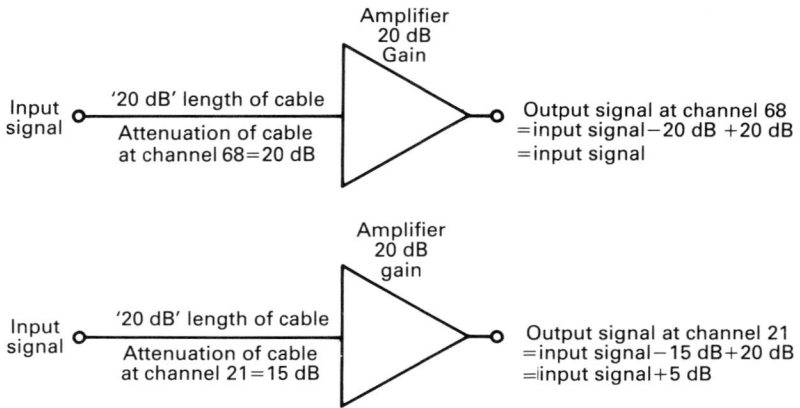

Fig. 5.17 — Showing how variations in signal levels occur owing to the attenuation of the cable changing with frequency.

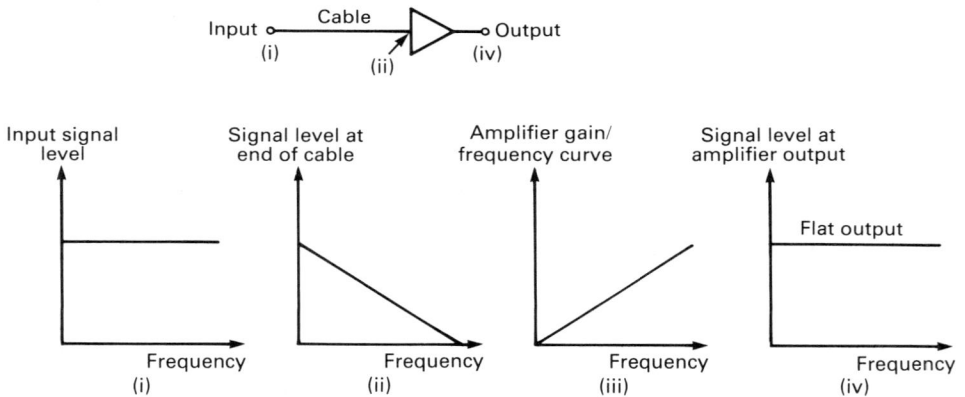

Fig. 5.18 — System with flat output.

known as its *slope*, and when the equalisation has taken place the system is said to be *flat*, in accordance with the shape of the amplifier output curve (iv). This *flat-output* mode of operation shown in the figure means that the levels of all the signals carried by the system, including the lowest and highest frequencies, are equal at the output of any amplifier.

Generally speaking the equalisation is achieved in the early stages of each amplifier, and one way to achieve this is for all signals to be passed through a passive frequency-conscious attenuator unit which reduces the levels of the lower-frequency signals to match those of the higher frequencies. The resulting low-level signal is then amplified by an amplifier with a flat gain/frequency characteristic. This arrangement is used in MATV systems serving blocks of flats or apartments, but is not so suitable for larger-scale CATV systems.

Unfortunately, connecting such a passive equaliser before the amplifier has the same effect as inserting an attenuator, in that the signal level at most frequencies will be reduced, whilst the noise will stay constant, so that the overall noise figure is increased, or, to put it another way, the signal-to-noise ratio is decreased (Fig. 5.19).

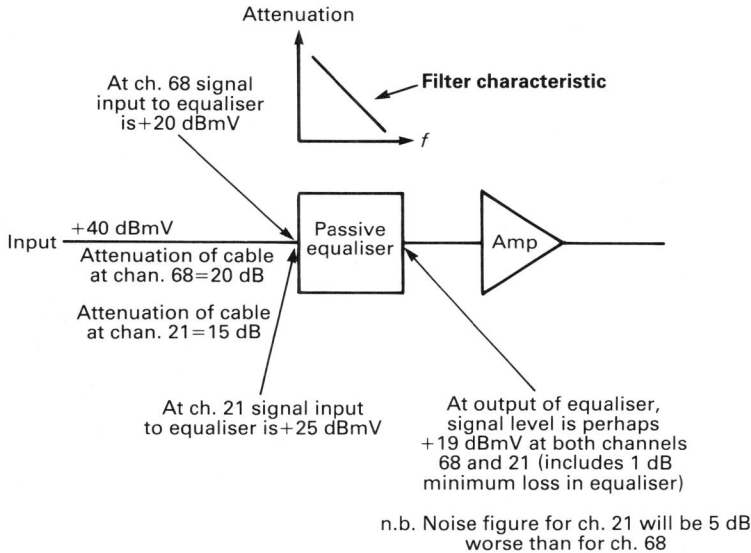

Fig. 5.19 — Variation of signal levels in system using passive equaliser before amplifier.

Since the difference between the output signals for the low-frequency and high-frequency channels in a system may be as much as 15 dB, this effectively means that the passive equaliser causes an increase in noise figure of 15 dB on some signals, which would generally be considered to be completely unacceptable since it would reduce the available system length significantly. This problem has led to an alternative way of obtaining equalisation, which is to design the amplifier itself with a gain characterisitic that increases with frequency. To design such a device so that it matches the frequency/attentuation loss of the system is difficult, but the advantages over the previously discussed idea are so great that this type of amplifier is the norm in large wideband systems.

Fig. 20 shows how such a system operates. The input to the system is arranged to rise with frequency so that the output at the end of a cable run is sensibly flat. Each amplifier is then designed so that its gain rises with frequency so that it mirrors the cable loss characteristic. Thus the signal levels applied to the input of the next section

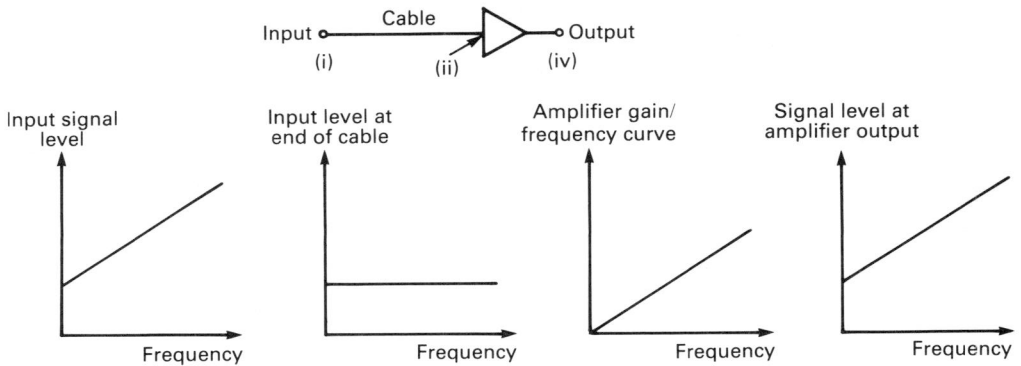

Fig. 5.20 — Signal-level variations in a system arranged for 'full-tilt' operation.

of the cable system are higher at the higher frequencies, so that by the time the signals have passed through the section they will be flat when they reach the input of the next amplifier. This method of operation is known as *full-tilt*.

An amplifier for use in a system of this type will have a 'tilt' control (should really be the 'slope' control to tie in with our earlier definition!), which will change the frequency response of the amplifier with gain, so that in an ideal system the loss of any length of cable could be compensated for in a manner which ensures that the necessary frequency/amplitude equalisation changes are correctly made. This control, also known as a 'tilt-compensated gain control', can be compared with the so-called 'loudness' controls found on some hi-fi audio amplifiers, which alter the frequency response of the amplifier as the volume is turned up or down. It is not possible to obtain perfect correction for all frequencies, and some amplifiers also have flat gain controls, but the interaction of the two can often make it difficult to obtain proper correction, and this type of equipment is deprecated.

Full-tilt operation is nowadays generally recommended, since the lower gain required at some frequencies can lead to lower distortion, but old systems sometimes encountered noise problems at the lower frequencies, which could be overcome by adopting the so-called *half-tilt* approach, which, as can be seen from Fig. 5.21, is a compromise between the 'flat' and 'full-tilt' arrangements.

5.2.2   Distortion reduction — coherent and harmonically related carriers

We have seen that the major forms of distortion that worry cable operators are second order and third order (cross-modulation), and that the number of spurious products increases rapidly as more channels are added, cross-modulation products being proportional to the square of the signal levels. Since the vision carriers have the highest signal levels, energy peaks occur during the synchronising pulses of the vision signals, and these can modulate the carriers that are producing the distortion. If the energy peaks are time coincident for several different carriers, a build-up can occur, which increases distortion significantly, unless signal levels are reduced throughout the system. If all channels are scrambled, or a technique known as 'all-channel sync suppression' is used, the energy peaks are removed, so that the energy is spread fairly

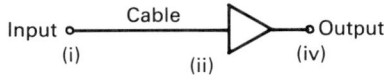

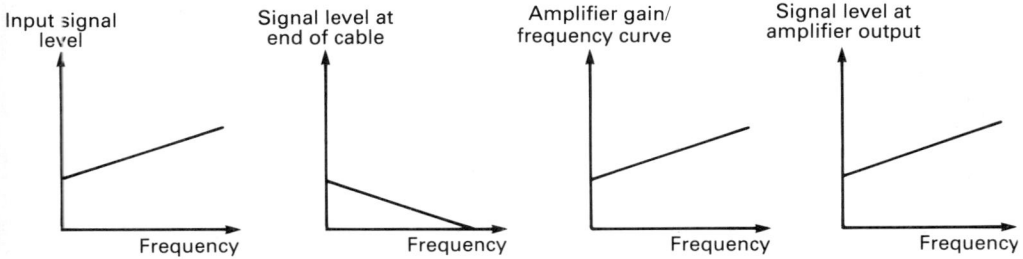

Fig. 5.21 — Half-tilt system.

evenly over the spectrum and interference due to the 'triple-beat' effect is very much reduced. Even this is not always enough, however, and scrambling of this type may not be possible in some sytems.

It has, however, been found [ref. 3] that if the carrier frequencies of the various signals are harmonically related and, ideally, phase coherent, the distortion, which normally shows up as various forms of patterning, is much reduced. This is because the intermodulation products of the various vision carriers are made to be phase coherent with the carriers, and the effects of these intermodulation products therefore disappear, leaving only the sidebands of the modulating signals to cause patterning. It is generally accepted that distortion levels can be reduced by about 10 dB using these techniques, allowing either for the system signal levels to be increased by 5 dB, or for more channels to be added to an already fully loaded system, without an increase in distortion. It is reasonably simple to apply this technique in most systems which use IF to channel convertors, by phase-locking each convertor to the outputs of a common 'comb' frequency generator.

REFERENCES

[1] W. Rheinfelder, Calculating overload thresholds for cascade amplifiers. Electronic equipment engineering. July 1964 pp. 63–65.
[2] V. Lewis. Television distribution basic planning guide. Wolsey Electronics Ltd.
[3] W. T. Homiller, Headend techniques for reducing distortion, Jerrold, Hatboro, Pa. USA, NTTA tech., 1983.

6

Amplifiers

6.1 TYPES OF AMPLIFIER

Our discussion of amplifiers in the distribution system so far has concentrated on the concept that an amplifier is inserted into the distribution system at various points in order to compensate for the effects of the losses that the signals sustain whilst passing through that section of the cable. This type of amplifier is generally known as a *trunk* or mainline amplifier, but other types are also required for other purposes at different parts of a system, as shown in Fig. 6.1. All the types are basically similar, and the foregoing remarks on noise, distortion and equalisation will generally apply, but these other amplifiers are intended for specialised purposes. It is common these days for amplifiers to be built on a modular basis, with sockets being provided for the optional addition of various filters, equalisers, bridger modules, etc., thus allowing the manufacturer to benefit from being able to make use of large numbers of basically identical amplifiers, whilst permitting the user to achieve flexibility by the selection of different modules to plug into different amplifiers.

The *bridger* amplifier (Fig. 6.2) is used at points where a branch is taken off the main trunk to serve a group of houses or a district. This provides enough gain to overcome the loss caused by its being put into the line (insertion loss), and usually has several outputs which are effectively isolated from the main line by its circuitry. Bridger amplifiers are often combined in the same case as the trunk amplifier, and are referred to as a mainline–bridger combination (Fig. 6.3). The outputs of the bridger amplifier are connected to the cables which are to be run down individual streets or to serve small groups of houses, and since the bridger outputs will not usually be passing through many other cascaded amplifiers it is usually found useful to increase the levels of the bridged outputs above those on the main line. This enables the signals to be run to the houses with the best possible signal-to-noise ratio, without having to worry about the problems of intermodulation distortion that would occur in a part of the system using many cascaded amplifiers.

The actual tapping off of the signals from the trunk cable is usually accomplished

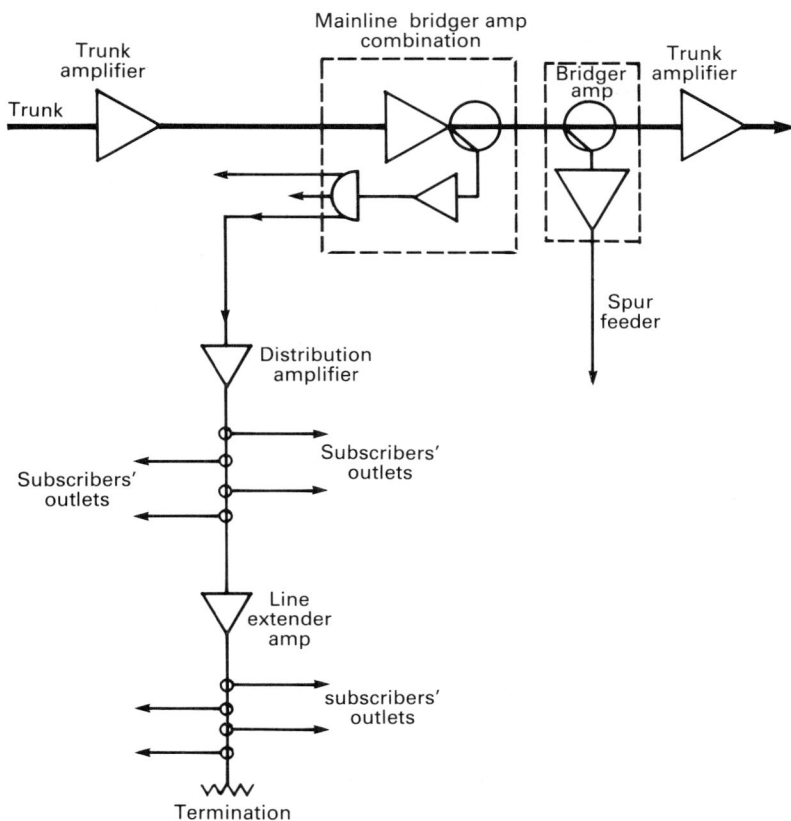

Fig. 6.1 — System diagram showing use of amplifier types.

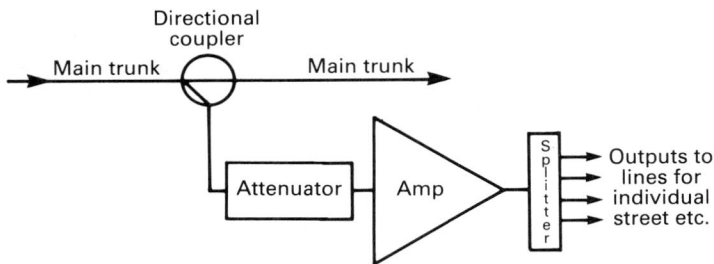

Fig. 6.2 — Basic bridger amplifier.

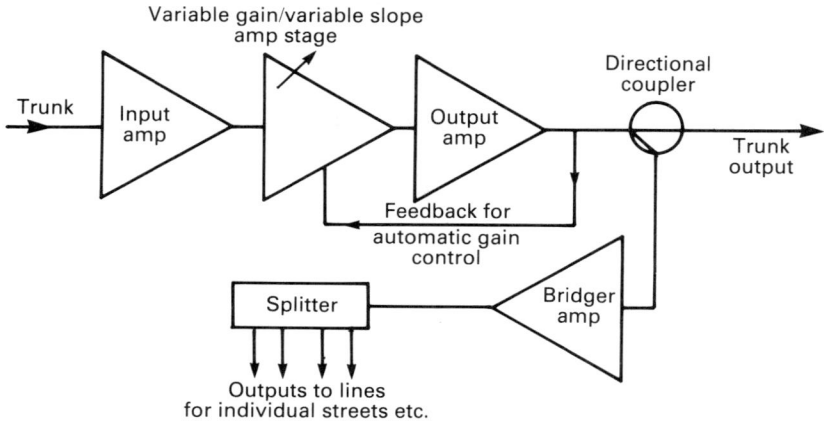

Fig. 6.3 — Mainline-bridger combination.

by a directional coupler, the operation of which was described earlier (Chapter 3), and the tapped-off signal is then amplified and split to give multiple outputs. The signal levels need not be the same at each output, if an asymmetrical system design is required. In modular design systems a bridger amplifier is obtained by plugging an appropriate module into a standard amplifier case.

Although we indicated above that the outputs from a bridger would not normally be passing through numerous cascaded amplifiers, it is sometimes necessary to increase the level of the signals in a distribution feeder, and this is done by a *line-extender* amplifier, or a distribution amplifier (Fig. 6.4). Again, line-extender

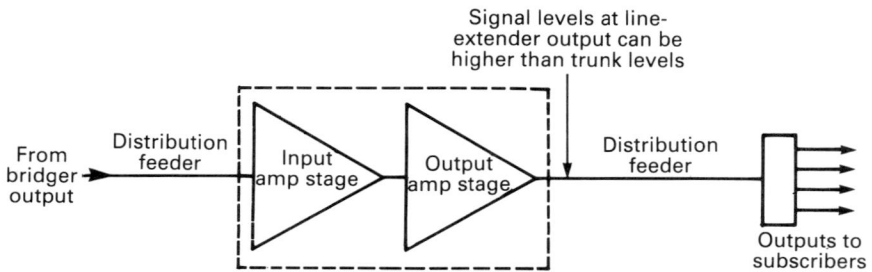

Fig. 6.4 — Line-extender amplifier.

amplifiers can have somewhat lower specifications than trunk amplifiers because they will not generally be used in cascade. The output of a line extender will sometimes be split into two or four, for feeding to subscribers. Line extenders often operate at higher output levels than trunk amplifiers in order to provide outputs to more subscribers economically, without distortion becoming a problem.

6.2   AMPLIFIER DESIGNS

Amplifier constructions that are commonly used are shown in Fig. 6.5, and it will be

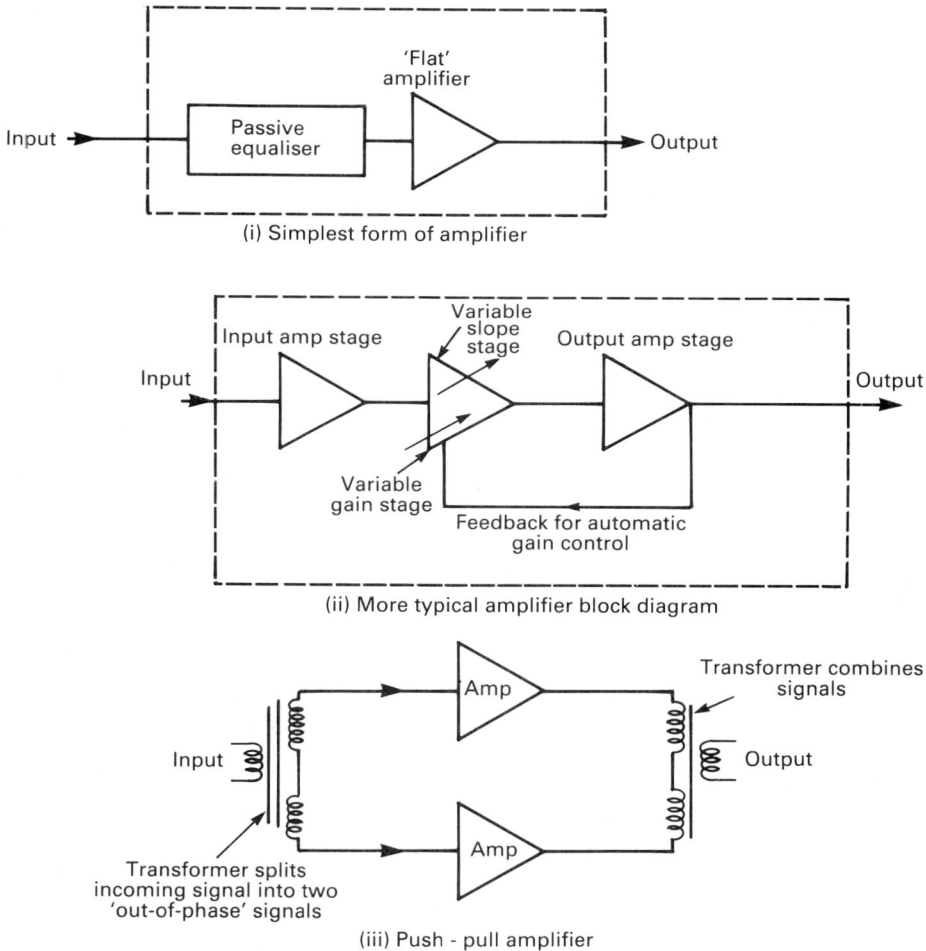

(i) Simplest form of amplifier

(ii) More typical amplifier block diagram

(iii) Push - pull amplifier

Fig. 6.5 — Basic amplifier designs.

remembered that we have already established that the use of the first type, (i)—a flat amplifier preceded by a passive equaliser, is not generally considered to be good practice.

Fig. 6.5 (ii) illustrates a more commonly encountered type of amplifier, where two or three separate stages of amplification are used, and feedback signals from the output stage are used for automatic gain control. The design of such an amplifier is usually arranged so that, as far as possible, the first stage is responsible for the noise performance of the amplifier, and the final stage determines the level at which overload will occur. The intermediate stage contains the equalisation circuitry, and sometimes both gain and slope controls are included to facilitate adjustments in the field, although such adjustments are generally frowned upon once the system levels have initially been determined. Far too frequently enthusiastic technicians with test equipment accurate to only + or − 6 dB try to adjust signal levels to the nearest dB, in order to overcome a problem at some point along the chain, with disastrous consequences to the rest of the system.

In our earlier discussions on types of distortion we saw that second-order-distortion products can easily become a nuisance in systems which carry many channels, since the spurious products are likely to fall on frequencies being used for programme carriers. Since the earliest days of electronics, so-called 'push–pull' audio amplifiers have been used to cancel out second-order distortion in equipment with 'high-fidelity' aspirations, and the same techniques are used in some cable amplifiers. Fig. 6.5(iii) shows how such a system works. The incoming signal is split into two signals which are 180° out of phase with each other; this can be achieved by a specially designed transformer or appropriate transistor circuitry. The two out-of-phase signals then pass through separate amplification stages, and are combined in an output transformer arranged so that the wanted signals add, whereas any second-order-distortion signals that are produced will cancel because they are effectively in phase as they are added to the two opposing windings of the output transformer.

6.2.1 Feedforward amplifiers

By designing and building fairly complex amplifier systems it has been found possible to achieve significant improvements in gain and distortion over conventional cable television amplifiers. The feedforward design came into use on wired distribution systems about 15 years ago, and what started off as a fairly expensive design intended for long-distance trunk routes has now become a standard amplifier type, which can be used for many applications where its extra gain and lower distortion can justify the increased cost.

Fig. 6.6 shows the basis of operation of a feedforward amplifier. It consists of two conventional integrated circuit (IC) amplifiers, four directional couplers, two delay lines which also act as inverters, and an attenuator. The first amplifier is the main amplification stage of the unit, and things are arranged so that the distortion products which are generated in this stage are fed forward, amplified and cancelled in the output stage, so that we are left, theoretically, with only the original signal in its amplified form.

The incoming signal is first split in the directional coupler DC1. The part of the signal passing through the delay line DL1 is then inverted and inserted into the directional coupler DC3; a small part of the signal that has passed through the main amplifier is tapped off by the directional coupler DC2, its level is adjusted by the attenuator to match that of the incoming signal to DC3, and when the inverted signal combines with the non-inverted one, the two cancel, so that the programme signals cancel out, leaving only the distortion products that have originated in the main

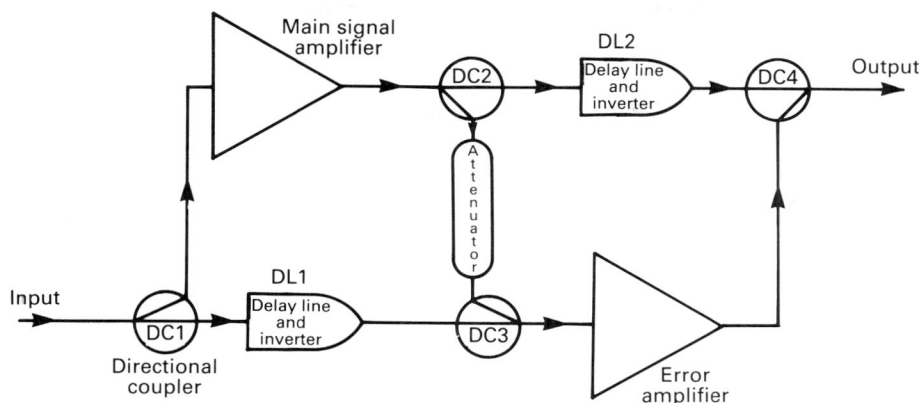

Fig. 6.6 — Main elements of a feedforward amplifier.

amplifier. This distortion signal is then amplified in the error amplifier and combined with the inverted signal from the main amplifier in directional coupler DC4. Provided that the gains of each stage and the signal delays have been properly set up, the distortion products will then cancel in DC4, leaving only the amplified programme signal. Complete cancellation can never be achieved in practical amplifiers, but in a typical amplifier module using IC amplifiers of about 35 dB gain, it is often possible to achieve perhaps 25 dB of cancellation.

The net result of all this is that it is possible to design an amplifier with a higher-than-normal gain, but with significantly lower distortion. It might also be thought that the noise performance would be improved, since most of the noise will effectively be part of the error signal that is cancelled. This is not true. It is not theoretically possible to cancel amplifier noise which is present in the input signal. In a feed-forward amplifier, in the ideal case, the error amplifier will have no noise signal on its output so that the noise of the main amplifier cannot be cancelled. In practice, however, it is found that the noise figure is about 3 dB worse than that of a standard amplifier because of the inevitable loss between the input terminal and the error amplifier.

6.2.2 Power-doubled amplifiers

Fig. 6.7 shows yet another amplification technique that is used in CATV systems. Originally introduced to increase system reliability and to provide a greater maximum output level than can be achieved with a simple amplifier design, the so-called power-doubled technology is nowadays often used to keep distortion products low, whilst achieving good noise performance.

The basis of the design is to use two conventional integrated circuit amplifier stages in parallel, with a splitter at the input, and a combiner at the output, each of which should theoretically introduce about 3 dB of loss. Thus the input signal to each

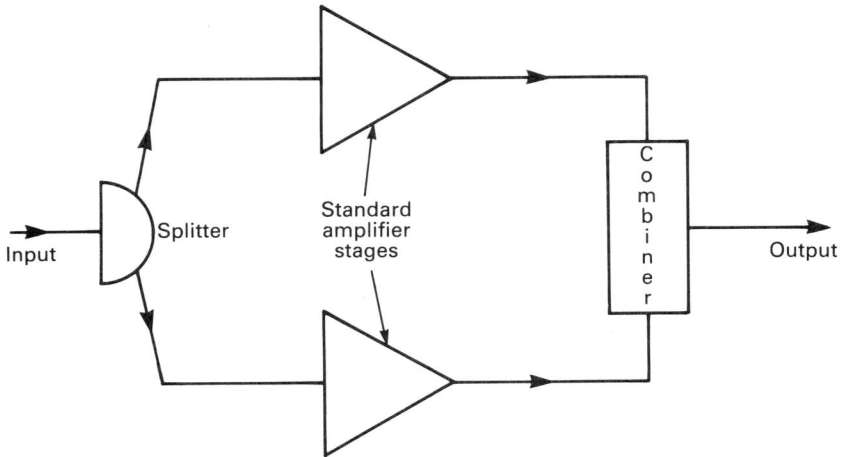

Fig. 6.7 — The elements of a power-doubled amplifier.

of the IC amplifier stages is 3 dB below the input signal to the actual amplifier. Each of the two parallel stages then amplifies the signal, and their outputs are combined, adding together in voltage terms, since both signals are in phase. The output signal level is therefore increased by 6 dB compared with the gain of one of the stages, but is reduced by 3 dB in the combiner. Since the input signal to each of the IC stages was also reduced by 3 dB because of the splitter, the net result of all this is that the power-doubled amplifier has exactly the same gain as one of the individual IC amplifiers. If we remember, though, that each IC is actually operating at 3 dB below its specified output level, we can see that the distortion will be reduced by twice this, i.e. 6 dB. Since the noise signals generated in each IC are quite independent of each other, they will be incoherent, and will not therefore add in the same way that the programme signals did. Thus the noise for the two amplifiers together is only the same as for one of the IC amplifiers.

So in theory the more complex power-doubled design of amplifier can give a 6 dB improvement in distortion, with the same gain and noise performance as a standard amplifier. In real life we find that the splitter and coupler will each have a loss of at least 3.5 dB, rather than the 3 dB our theory assumed. This means that the gain of the stage will be reduced by 1 dB altogether, and that the noise will be increased by half a decibel. We therefore find that the distortion figure is actually improved by 5 dB, which is still a very worthwhile increase, but the cost of this is that the power consumption is doubled and the complexity and therefore the cost of the amplifier are increased. One other advantage of this type of design is that clipping of the signals will not start to occur until the output signal is about 2 dB higher than that normally required to cause this to happen. A look at Fig. 6.7 shows that the power-doubled design also gives some output even if one of the IC amplifiers dies, which improves overall system reliability, even though the signal level will be reduced by 6 dB and the carrier-to-noise ratio will drop by about 3 dB.

6.3 AUTOMATIC GAIN/LEVEL CONTROL — AUTOMATIC SLOPE CONTROL

In Chapter 3, dealing with the head-end equipment, we mentioned the need for some form of automatic gain control (AGC), and indicated that pilot carriers at frequencies close to the highest and lowest frequencies in use could be used to achieve AGC. If the level of either of these incoming pilot signals should rise or fall for any reason, the gain of the amplifier will be decreased or increased appropriately, in order to keep the output level sensibly constant at each frequency. It should, however, be noted that major changes cannot be achieved without a corresponding decrease in signal-to-noise ratio or an increase in cross-modulation, so AGC cannot ever be regarded as a substitute for proper initial setting up of system levels. Just one pilot carrier can be used, as was the case in many older systems, but these days it is normal to use two, so that as well as AGC, automatic slope control (ASC), i.e. control of the gain of the amplifier at two particular frequencies, is also achieved. Some engineers also distinguish between the terms 'automatic gain control' and 'automatic level control' (ALC), using AGC to describe the automatic control achieved by using the level of one of the television signals being carried as the control signal, and ALC when discrete pilot signals are used for this purpose.

  Although variations in the incoming signal at the head end may occur, and amplifier gains may well change as the equipment ages, or due to faults, the major cause of signal-level variations in a cable distribution system is changes in the ambient temperature. Any amplifier will be affected to some extent, but these days modern transistorised equipment can be designed to achieve a variation of no more than a quarter of a dB over a temperature range of -45–$+45°C$, which is insignificant. The main thermal problem is invariably due to the fact that cable losses change significantly with temperature variations. A moderately good cable might have a specified temperature variation of 0.002 dB per dB of length, per degree Centigrade. A system with a total cable length of 1000 m, perhaps 70 dB in cable attentuation terms, might therefore have a variation of

$$0.002 \times 70 \times 50 \,°C \,(-15 \text{ to } +35, \text{ say})$$
$$= 3.5 \text{ dB}$$

which would have a significant effect on system performance unless corrected for. Although it would appear to be possible to compensate for these temperature changes by installing thermistor-type devices at appropriate parts of the system, in practice it is found to be impossible to make such a system work satisfactorily; invariably the temperature sensing and control circuitry is in a box at a different temperature from the exposed cables that it is supposed to be compensating for. An AGC system, which tries to keep the output of an amplifier constant for variations in the input level, whether these have been caused by standard cable losses or temperature variations, is a far more satisfactory way of achieving the required compensation, and is used in nearly all modern systems. It is also usual for the AGC to be frequency–conscious, in order to automatically compensate for the temperature variations at each signal frequency, so that once again the automatic slope (or tilt) control, sometimes described as a tilt-compensated gain control, is used to keep signal levels constant.

It is worth noting that the variation in temperature of a cable buried underground will be very much less than that of a similar cable strung from overhead poles. Underground cave systems in the UK tend to have a fairly constant mean temperature of about 10 °C throughout the year, and it has been found that cables buried at a depth of about 75 cm also have a year-round average temperature of about 10°C, although variations are obviously greater than in a deep cave!

The addition of automatic gain or slope control to an amplifier increases its cost, which has led to many studies being carried out as to how frequently AGC amplifiers need to be used. Higher-gain amplifiers, which minimise costs because they are required to be installed only at long intervals along the distribution system, will generally give rise to the need for large amounts of AGC, since signal levels can vary significantly as they pass through long lengths of cable, and in practice it is found that trying to correct for losses using large amounts of AGC leads to poor performance at the end of a long system. Rule-of-thumb methods of accounting led to many systems using AGC at every third amplifier, although modern transistorised equipment has reduced the relative cost of AGC modules, so that it is currently fashionable to put AGC on alternate amplifiers throughout the chain. Even this can lead to problems, however, when a trunk amplifier without AGC is used to feed several distribution amplifiers, since subscribers near the end of the line will be receiving their signals through two or even three unregulated amplifiers in cascade. To overcome this type of problem some systems choose to have AGC on all their trunk amplifiers, and other systems use strategically placed AGC amplifiers to prevent isolated customers on the end of long distribution lines from suffering from large variations in level.

The latest amplifier designs, which are based on the use of integrated circuit technology, can incorporate more sophisticated circuitry without a significant price increase, because of the mass-production methods that are used in their manufacture. It seems likely that the extra cost of adding AGC and ASC to such designs will become negligible, which will make it possible for future cable systems to have AGC and ASC on all their amplifiers, which will lead to much closer control of levels throughout the system.

6.4 POWER SUPPLY METHODS

In an ideal system it would be possible to position each amplifier in a building which had its own main supplies, so that each amplifier could be individually connected to the mains. This is hardly ever possible, however, in any practical system, where some amplifiers will be mounted on poles or in other isolated or difficult-to-reach positions, so that a good deal of thought has to be given to providing power for each amplifier. Old systems using valve amplifiers which consumed a great deal of power often had no alternative but to power each amplifier separately, and sometimes this led to amplifiers being spaced as far apart as possible to minimise cost and installation difficulties, with the consequence that the signal quality was often much worse than it should have been. Nowadays, when relatively low-power transistor amplifiers are the norm, it is usual to feed the required power along the same distribution cable as the signal, providing power inputs at the beginning of the system and also at other convenient points along the route, where amplifiers and the cable can be positioned within easy reach of a mains supply. Various precautions have to be taken to ensure that the power can be transferred along the cable safety and without causing degradation to the television and sound signals.

Direct current (DC) is invariably required for amplifiers, and some early systems passed DC along the cables. It was soon found, however, that the passage of large direct currents through the various dissimilar metals that make up the cables, terminals and connectors of a wired distribution system caused corrosion due to electrolytic action, especially in locations where it was difficult to completely exclude atmospheric moisture. This led to the modern practice of alternating current (AC) supplies being fed along the cable, with the rectification, i.e. conversion to DC, taking place in the power supply unit. For safety reasons the alternating mains voltage is reduced from 100 to about 60 V at 60 Hz in the USA, and from 240 to around 30 V at 50 Hz in the UK, and this alternating supply is then coupled into the cable by means of an inserter, which operates as shown below. British Standard 6513:3:3:5:3 specifies that the voltage between inner and outer of the distribution cable must not exceed 63 V AC(rms), which is equivalent to 90 V peak. The corresponding international standard IEC 728:65:3:1 specifies that this voltage must not exceed 65 V rms.

6.4.1 Power inserters

The inserter must appear to the radio-frequency signals as though it is just another length of coaxial cable, and must therefore be carefully impedance matched, so that its insertion does not cause any reflections due to mismatches, which could show up as 'ghost' images on the pictures or lead to corruption of teletext displays. As long as these conditions are satisfied, a power inserter can be put into the cable at virtually any point required. A typical inserter might have an insertion loss of perhaps 0.5–1.0 dB. The inserter can be a completely separate unit that is merely plugged in series with the distribution line, or it can be included as part of the power supply, or combined with an amplifier (Fig. 6.8).

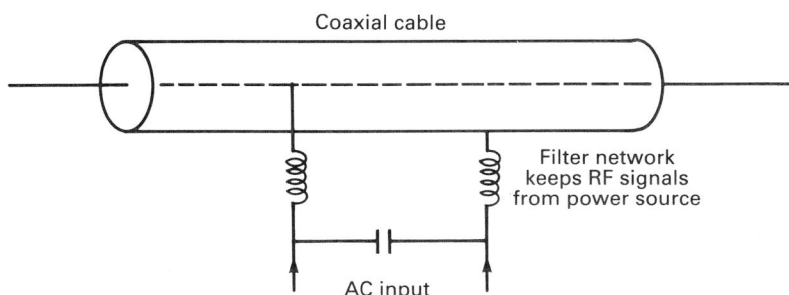

Fig. 6.8 — Method of coupling AC power into the cable system using an inserter.

It will be seen that the network of capacitors and inductors is arranged so that the low-frequency current from the power source is kept away from the signal path, so that noise and mains-borne interference are kept to a minimum, and so that there is no chance of the radio-frequency signal being shorted out through the power supply.

Since it is AC power that is fed to the line, and the amplifiers will need DC, it is usual to rectify the AC in a power supply unit (PSU) forming part of the inserter, and Fig. 6.9 shows how the DC power is fed to the amplifier, whilst the low-frequency AC

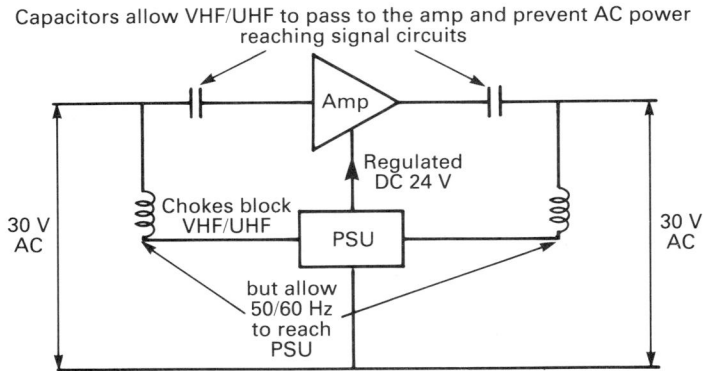

Fig. 6.9 — How an amplifier obtains its DC power, taking its AC power from the wired distribution system.

is carried along the line.

Even with this type of circuit, a substantial DC component can flow along the cable if simple half-wave rectification is used in the PSU, and corrosion problems found only after long experience led to the practice of full-wave or bridge rectification being used in all modern systems. It is usual to make use of regulated power supplies to ensure that the voltages fed to the system remain sensibly constant even though the mains voltage varies. Note that the chokes block the radio-frequency signals, preventing their passing to earth via the PSU, whereas they allow the low-frequency power signals to pass through to the PSU. In a similar way, the capacitors are of such value that they stop the mains-frequency power from reaching the signal circuits of the amplifier, whilst allowing the radio-frequency television signals to pass through the amplifier.

Switchers
Many modern power supply units are designed on the switched-mode power conversion principle, since this type is amongst the most efficient, and for this reason power supply units are sometimes known, rather confusingly, perhaps, as *switchers*.

6.4.2   Powering modes
Individual amplifiers can be powered in various ways. If an amplifier is 'input powered', power from the power supply is fed to the amplifier through its input connector, and as far as power is concerned, the output of the amplifier may be isolated from its input. Any power that happens to be present at the output port could come from a different power supply unit further along the distribution cable. Sometimes the power for an amplifier is obtained from the output connector, and

sometimes the 'through-powered' condition is used, where the outputs and inputs of the various amplifiers are effectively connected together. This has the advantage that an amplifier can obtain its power from either upstream or downstream, which can improve the reliability of a system, since in the event of the failure of one power supply, an amplifier can take its power from another power supply unit which is connected elsewhere on the line.

6.4.3   Inserter spacing — loop resistance

We have already seen how important it is to design systems so that signal losses are kept to a minimum, and now we must apply the same type of reasoning to the power supplies. Whereas radio-frequency signals pass through a coaxial cable with a particular loss which is dependent on the capacitance and inductance of the cable, as well as its resistance, the low-frequency power signals will be affected mainly by the resistance of the cable, and as the power signals are effectively passing through both the inner and the outer of the cable in series, it is this so-called 'loop resistance' that must be used when making calculations for power supplies. Fig. 6.10 shows the

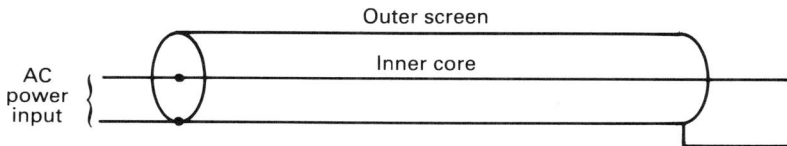

Fig. 6.10 — 'Loop resistance'.

meaning of 'loop resistance', and it will be seen that if a particular cable had a loop resistance of perhaps 3 Ω/km, a current of 10 A, which is typical of what might be obtainable from a practical inserter, would cause a voltage drop of 15 V to occur every half-kilometre along the cable. Since the system designer must ensure that each amplifier is presented with at least its minimum specified operating voltage, this knowledge will make it possible to calculate the minimum spacing for power inserters.

6.4.4   Standby-power arrangements

The vast majority of systems obtain their power from the AC mains supply, and in the UK the reliability of the mains is generally reckoned to be so high that alternative supplies are not usually provided. The argument put forward is that if the mains supply to the equipment serving a reasonable-sized system fails, then the supply to the customers' homes will also have failed, so that they would not be able to watch television anyway. In other parts of the world, the reliability of the electricity mains can leave something to be desired, and on very large systems it would be perfectly possible for viewers to lose the incoming wired services even though their own mains electricity supply was present. For these reasons it is often necessary to consider the provision of standby-power arrangements. We have already seen that through-powering of amplifiers gives rise to a certain amount of protection, since if one power supply unit fails, the amplifier may be able to draw its supplies from further up the cable. In practical systems it is often found that the loop resistances are such that a power inserter is needed for every six or eight amplifiers.

At critical points in the network it is possible to install standby batteries which are automatically recharged when the mains supply is present, or batteries which power an inverter to provide a temporary replacement for the mains supply to the inserter. It is not usually considered worth while to install motor-driven generating equipment, since this can prove expensive, requires a good deal of maintenance, and the overall reliability of auto-start motor-generating sets is often found to be poorer than that of the mains supply they are protecting. This has led to situations where the generator was only called upon to work once in a couple of years, and when it was needed the auto-start system failed.

7

Matching, reflections and echoes

As we have considered the various parts of a cabled distribution system, we have seen the importance of ensuring that noise and distortion are kept to as low a level as possible, and that variations in the amplitude/frequency response of the system must be minimised or corrected for. On several occasions we have also briefly mentioned the importance of keeping reflected signals to a low level, and have stressed the importance of correct matching when components such as taps and power inserters are being connected into the system. We shall now take a slightly more 'in-depth' look at the theory and practice of matching in order to try to understand the importance of this subject to a properly engineered cabled distribution system.

7.1 THE CABLE NETWORK AS A TRANSMISSION LINE

The cable forming part of a distribution system can be considered as a specialised case of the general transmission line, whose theory has been well understood for many years. A transmission line has four main parameters per unit length, resistance R, capacitance C, inductance L, and shunt conductance G, and can be represented by the equivalent circuit shown in Fig. 7.1. Obviously the various parameters shown

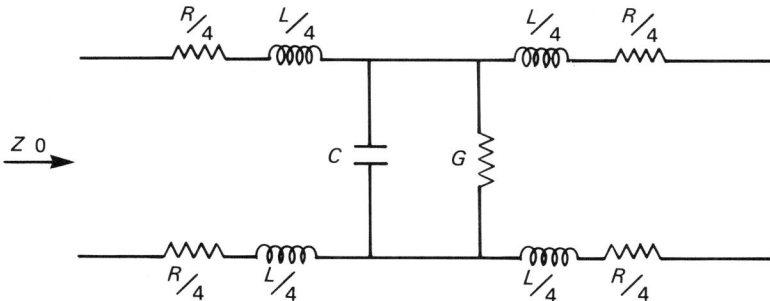

Fig. 7.1 — Equivalent circuit of a transmission line showing how the various parameters can be represented by 'lumped constants' L, C, R, G, where L = inductance per unit length, C = capacitance per unit length, R = resistance per unit length, and G = shunt conductance per unit length.

are distributed throughout the length of a practical cable, but it is sometimes helpful to think that with a sufficiently short piece of line they could be represented by the lumped constants shown as L,C,R,G.

The impedance seen when looking into the input terminals of a very long line is known as the characteristic impedance of the line Z_0, and is given by the expression

$$Z_0 = \sqrt{\frac{(R + j\omega L)}{(G + j\omega C)}}$$

where $\omega = 2\pi \times$ frequency. Since the resistance R in ohms and the conductance G in ohms are generally very small numbers in comparison with $j\omega L$ and $j\omega C$ for most cable systems, it is reasonable to disregard them when trying to simplify the situation for purposes of explanation, and it is commonly stated, therefore, that in a loss-free situation the characteristic impedance is given by:

$$Z_0 = \sqrt{L/C}$$

Note that this term will only be a number, with no frequency-dependent components, and it can therefore be considered as a resistance. If we terminate a practical length of line with a resistance equal to the characteristic impedance of the line, all the signal energy flowing along the line (the so-called 'forward wave') will be dissipated in the resistance, and no energy will be left to be reflected back towards the start of the line. If however, the line is terminated in some other value of resistance, part of the energy will be reflected back towards the source (the 'reflected wave'), and the amplitude of the signal at any point on the cable system will be the algebraic sum of the forward and reflected waves. Since the frequencies of the forward and reflected waves are the same (they result from the same input signal), the net signal will vary in amplitude from a minimum value, E_{min}, at the points (nodes) where the two waves cancel each other, to a maximum value E_{max}, where the two waves add in phase. Thus the forward and reflected waves combine to form stationary or standing waves, and the signal levels, voltage or current, on a mismatched line will be dependent upon whereabouts on the line they are measured (Fig. 7.2(a)–(c)).

Various terms are used to indicate how well a line is matched, but it is important to realise that they are merely different ways of expressing the same thing, and so are very closely related to each other.

7.1.1 Standing-wave ratio
The simplest is probably the VSWR, voltage standing-wave ratio, defined as

$$\text{VSWR} = E_{max}/E_{min}$$

i.e. the ratio of the maximum signal voltage along the line to the minimum. As an example of the sort of figure to be aimed for, British Standard BS6330 suggests that in general the VSWR should not exceed 1.5:1 over the working bandwidth, but also

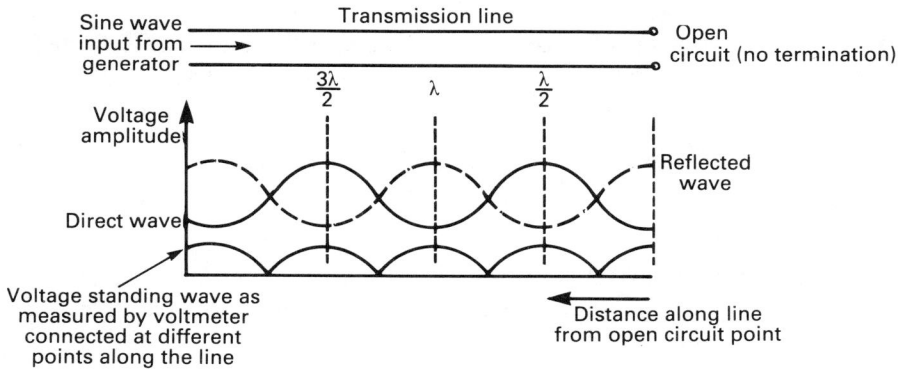

Fig. 7.2(a) — Showing how standing waves are formed from a combination of direct and reflected waves.

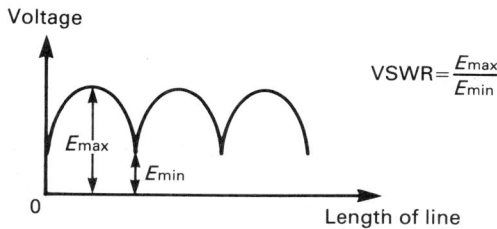

$$VSWR = \frac{E_{max}}{E_{min}}$$

Fig. 7.2(b) — Voltage standing waves on a line not terminated with its characteristic impedance.

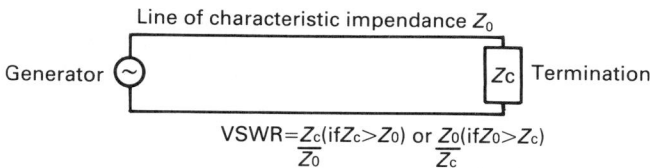

$$VSWR = \frac{Z_c}{Z_0}(if Z_c > Z_0) \text{ or } \frac{Z_0}{Z_c}(if Z_0 > Z_c)$$

Fig. 7.2(c) — Relationship between VSWR and termination impedance.

makes it plain that this limit, adequate for television pictures, may not be good enough for teletext-type data signal transmissions.

It should be noted that the VSWR is directly related to the impedance that is seen at the point where mismatching occurs, Z_c, by the formula

$$VSWR = Z_c/Z_0 \quad \text{or} \quad Z_0/Z_c$$

depending upon whether Z_c is higher or lower than the characteristic impedance Z_0 of the line.

7.1.2   Reflection coefficient

Another term is the reflection coefficient, defined as the ratio of the reflected voltage E_r to that of the forward voltage E_f (sometimes called the incident voltage):

$$r = E_r/E_f$$

7.1.3   Return loss

In practice, cable television engineers use yet another term, return loss, which is merely the value of the reflection coefficient r expressed in decibels, in order to make practical calculations simpler.

Return loss R is defined as

$$R = -20 \log r$$

which is the same as

$$R = 20 \log 1/r$$

Another way of defining the term 'return loss' is to say that it is the difference between the forward signal and the reflected signal (caused by a mismatch) along a cable, expressed in decibels. To see the usefulness of the term 'return loss', consider what happens when a piece of equipment such as an amplifier or a tap is inserted into a cable network (Fig. 7.3). Any practical piece of equipment will introduce some

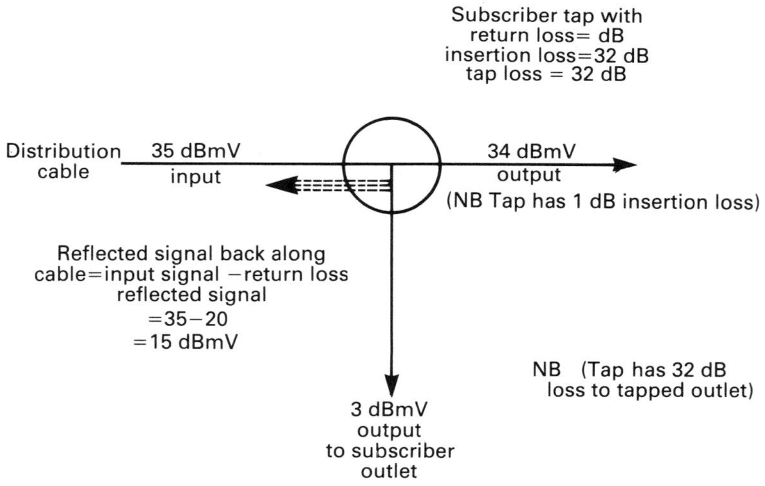

Subscriber tap with
return loss= dB
insertion loss=32 dB
tap loss = 32 dB

Distribution   35 dBmV   34 dBmV
cable   input   output
                                                           (NB Tap has 1 dB insertion loss)

Reflected signal back along
cable=input signal −return loss
reflected signal
=35−20
=15 dBmV

NB   (Tap has 32 dB
loss to tapped outlet)

3 dBmV
output
to subscriber
outlet

Fig. 7.3 — Illustrating the return loss.

degree of mismatch, and our particular device is quoted as having a 20 dB return loss in the manufacturer's specification.

If a 35 dBmV signal is present on the cable at the input to the device, the 20 dB return loss figure means that a reflected signal of $35 - 20$ dBmV, i.e. 15 dBmV, will be sent back along the cable towards the source. Therefore we can see that the higher the return loss of any piece of equipment, the less signal will be reflected back towards the source, and the better (lower) will be the reflection coefficient.

7.2 RETURN-LOSS FIGURES FOR PRACTICAL COMPONENTS

The International Electrotechnical Commission Standard, Publication 728, section 3.51, defines a test system as 'well-matched' when the ports facing the equipment under test have a return-loss ratio of at least 20 dB relative to the system impedance. Although figures for the return loss of the various items making up a cabled distribution system vary somewhat, as examples we might say that 14–20 dB is of the right order for a tap; a typical amplifier might have a return loss of 15–20 dB, although a low-cost VHF/UHF unit intended for MATV systems might be as poor as 10 dB. Customers' outlet sockets often have return-loss figures of around 13 dB, and coaxial cable invariably exhibits high figures, 30–40 dB being typical for a good-quality cable. It might be wondered why coaxial cable causes any reflections at all, but these come about as a result of internal mismatches. During manufacture it is found that slight irregularities in the concentricity of the rotating machinery used to make the cable cause slight variations in the diameter of the conductors and the spacing between them, and that these irregularities give rise to slight mismatching and reflections. This structural return loss of a cable can vary considerably with frequency, and wherever we talk of return loss we always mean the worst-case loss. British Standard BS 5425 specifies that the worst condition for trunk cables carrying signals from 5 to 450 MHz shall be a return-loss ratio of 20 dB, and that 18 dB is the worst permissible figure when frequencies from 450 to 1000 MHz are used. It should be realised, though, that the actual figure at most frequencies is likely to be very much better than this, as indicated above. It is difficult to accurately compare return-loss figures quoted for American and British cables, since two different methods are sometimes used to make the measurements, and the American results can often appear to be 6 dB different from those obtained using the technique specified in British Standard BS 5425.

It should be noted that if the cable were to be either open-circuited or short circuited, all the forward energy would be reflected back down the cable, from that point, so that the return loss would be 0%, and the VSWR would be $E_{max}/0$, which is infinity. This fact can be used to advantage when trying to trace the position of open- or short-circuit faults in a cable.

7.2.1 Relationships between terms
Fig. 7.4–7.7 illustrate how the terms voltage standing wave ratio, reflection coefficient, and return loss are inter-related.

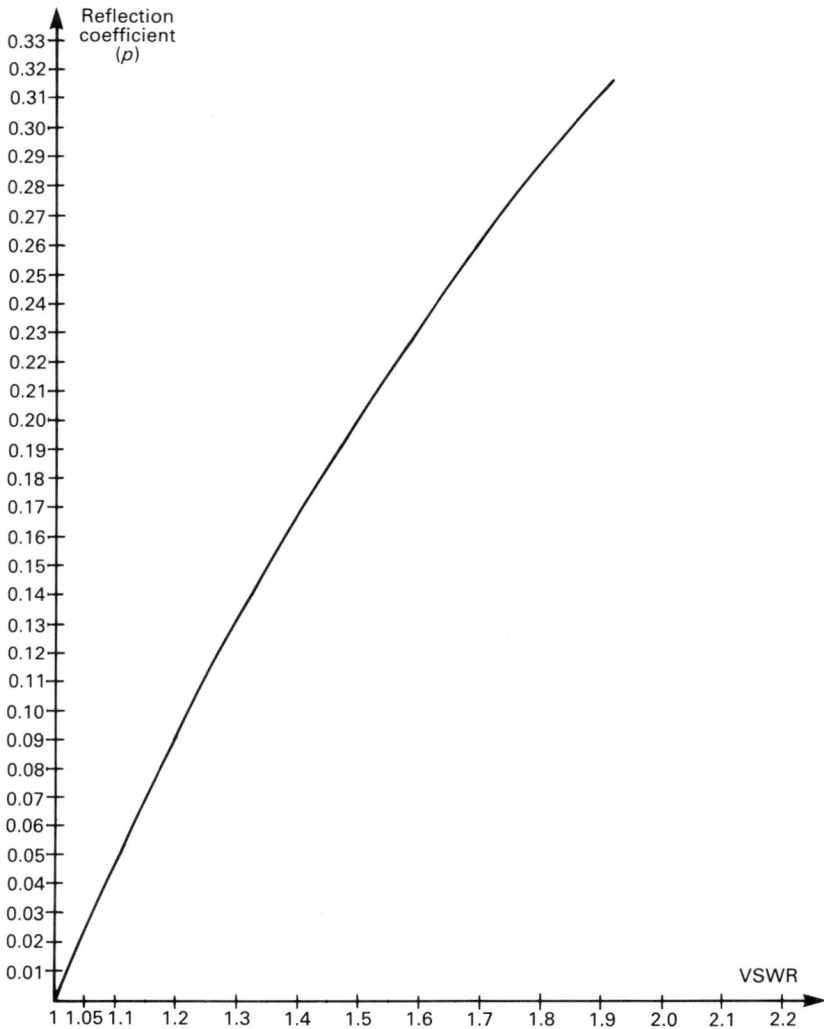

Fig. 7.4 — Reflection coefficient and VSWR (Courtesy IBA).

7.3   EFFECTS OF MISMATCHING

Even in the best designed practical cabled distribution system it is impossible to avoid the fact that there must be some degree of mismatching as device is connected to device, or even as an input signal to an amplifier passes from the coaxial input socket to be connected to the internal circuitry of the amplifier. As we have seen, any mismatch will give rise to some signal being reflected back along the cable from the point at which the mismatch occurs. If the reflected signals exceed a certain amplitude, they become visible to the viewer as 'ghosts', which look rather like those

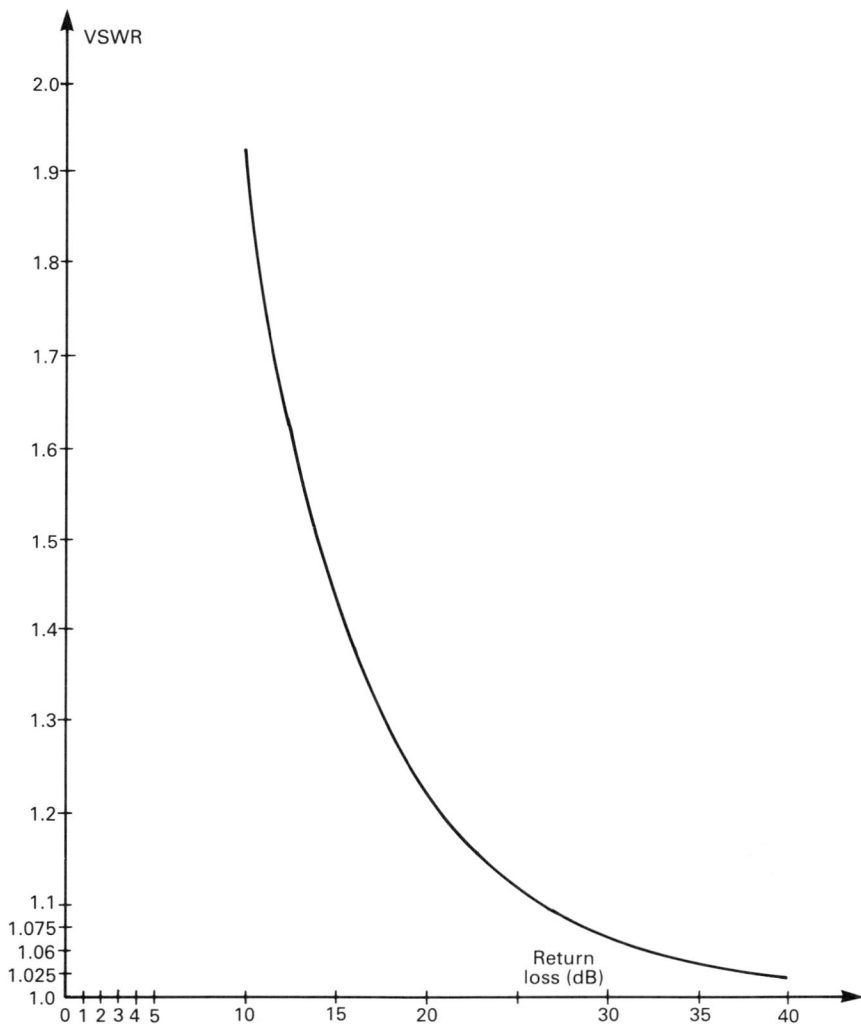

Fig. 7.5 — VSWR and return loss (Courtesy IBA).

caused when the user of an aerial system suffers from 'multipath' interference, the viewer seeing both the direct ray and the reflected ray, which has to travel over a longer path between transmitter and receiver and therefore arrives a few microseconds later than the direct image. The net result is that the viewer sees two images, one slightly displaced to the side of the other.

On cabled systems, whether or not such ghosts will be seen depends upon the amplitude of the reflected signal compared with that of the direct signal, and the time

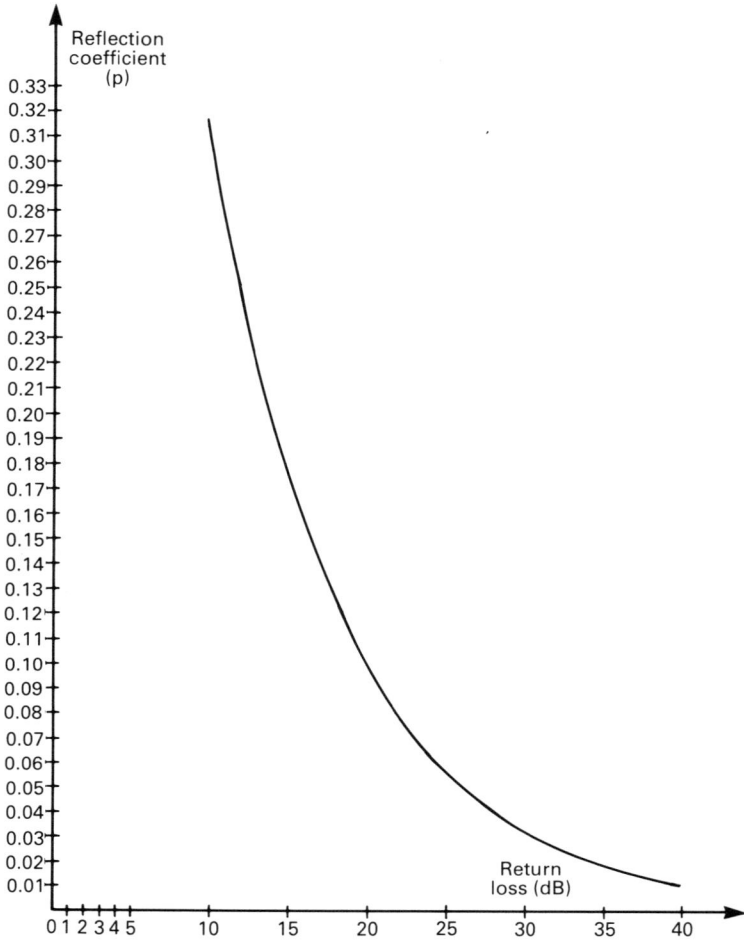

Fig. 7.6 — Reflection coefficient and return loss (Courtesy IBA).

difference between the two, which corresponds to the horizontal physical displacement. Visibility also depends upon the relative phase of the various reflected signals,
which tends to have a direct relationship with the spacings at which the mismatches
occur. If the physical spacing of the amplifiers, taps, etc., along the system coincides
with a half-wavelength, or multiples of half a wavelength at some frequency which is
being used, and it is difficult to avoid this occurring at some frequency when multi-
channel systems are in use, when reflected signals from one mismatch point travel
down the cable and add in phase with the reflected signal from the next mismatch
point, causing a gradual increase in the level of the 'ghost' signal. In theory the
opposite can also happen, with reflected signals which occur at quarter-wavelength

Nomogram 1

VSWR	Voltage reflection coefficient	Power reflected (%)	Return loss (dB)

```
Nomogram 1
                Voltage      Power      Return
                reflection   reflected  loss
 VSWR           coefficient  (%)        (dB)
  1.0            0            0          0
  1.1            0.05         0          25    24
  1.15                                   23    22
  1.2            0.10         1          21
                                         20
                                         19
  1.3                                    18
                 0.15         2          17
                                         16
  1.4                         3          15
                                         14.5
  1.5            0.20         4          14.0
                                         13.5
                              5          13.0
  1.6
                                         12.0
                              6          11.5
                 0.25
  1.7                         7          11.0
                              8          10.0
  1.8
                 0.30         9          10.5
  1.9
                              10         00
```

```
Nomogram 2
                Voltage      Power      Return
                reflection   reflected  loss
 VSWR           conffificient(%)        (dB)
  1.01           0.005                   40    45
  1.02           0.010        0          35
  1.03           0.015                   32    33
  1.04           0.020                   31
  1.05           0.025                   30
  1.06           0.030        0.1        29
  1.07           0.035                   28
  1.08           0.040                   
  1.09           0.045        0.2        27
  1.10           0.050                   26
  1.11                                   
  1.12           0.055        0.3        25
  1.13           0.060                   
  1.14           0.065        0.4        24
  1.15           0.070        0.5        23
  1.16           0.075                   
  1.17           0.080        0.6        22
  1.18           0.085        0.7        
  1.19                                   
  1.20           0.090        0.8        21
  1.21           0.095        0.9        
  1.22           0.100        1.0        20
```

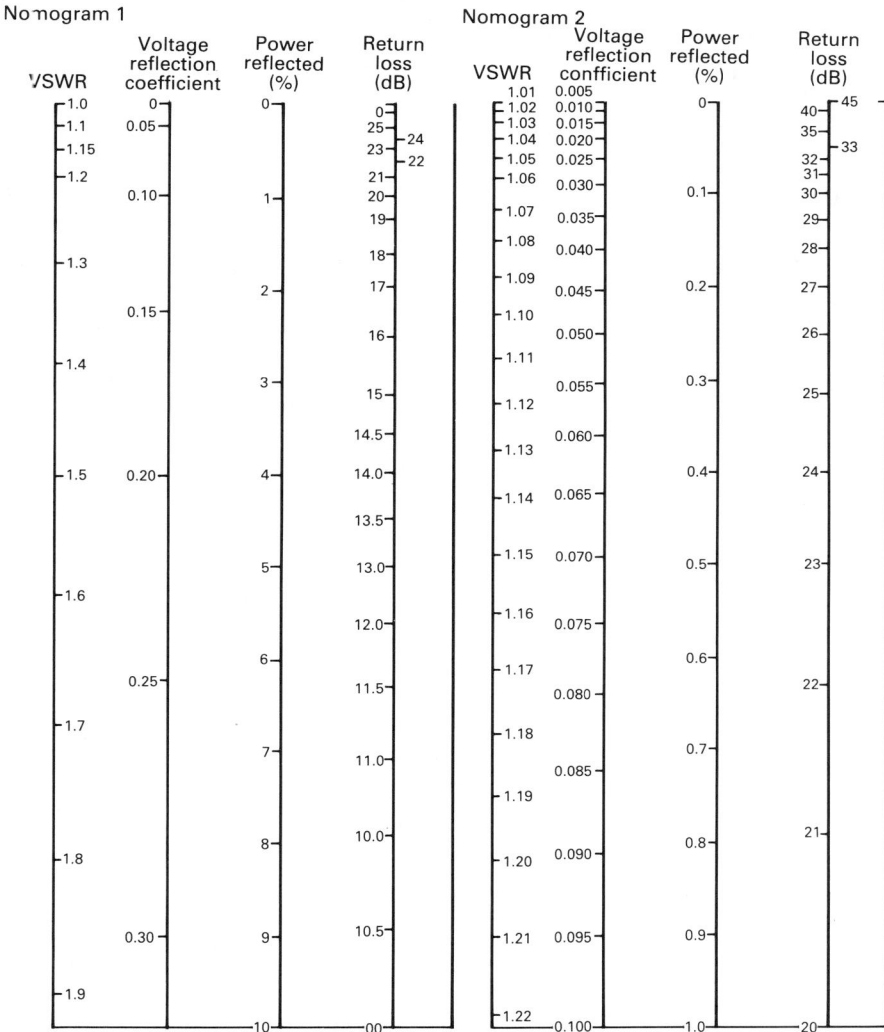

Fig. 7.7 — Nomograms showing relationship between VSWR, reflection coefficient and return loss (Courtesy IBA).

intervals cancelling each other out completely, but in practice it is not possible to take into account all possible mismatches when designing a system, so that this technique could not be relied upon to give a reflection-free system. It is worth noting, however, that reflections are generally less troublesome at the higher frequencies, where the amplitude of the reflected signal is attenuated by the cable to a much greater extent than at the lower frequencies.

A distinction is sometimes made between the terms 'echo' and 'ghost', or 'reflection', although many cable engineers use the words as though they were synonymous. Strictly speaking, any reflected signal that is fairly narrow-band in character, so that it only affects a small part of the frequency spectrum of any sound and television programme signal, perhaps the sound carrier, or the colour subcarrier, is called an 'echo'. If the reflected signal is wide-band, consisting of virtually all the frequency band used by a television signal, then it is known as a 'reflection', or a 'ghost', because the displaced image looks like the ghost or shadow of the main image.

Since the subjective effect of ghosting depends on so many factors, many attempts have been made to come up with strict limits for the amplitude of the reflected signal that can be tolerated for any particular time displacement of the image, since if the signal amplitude is small enough, no ghost will be visible, and if the displacement is very small, the effect on the picture may well be negligible. This work, involving many carefully controlled subjective viewing tests, led to various curves being drawn up to define the conditions under which ghosting would be perceptible, and some of these are shown in Fig. 7.8 [1]. Nearly 40 years after this

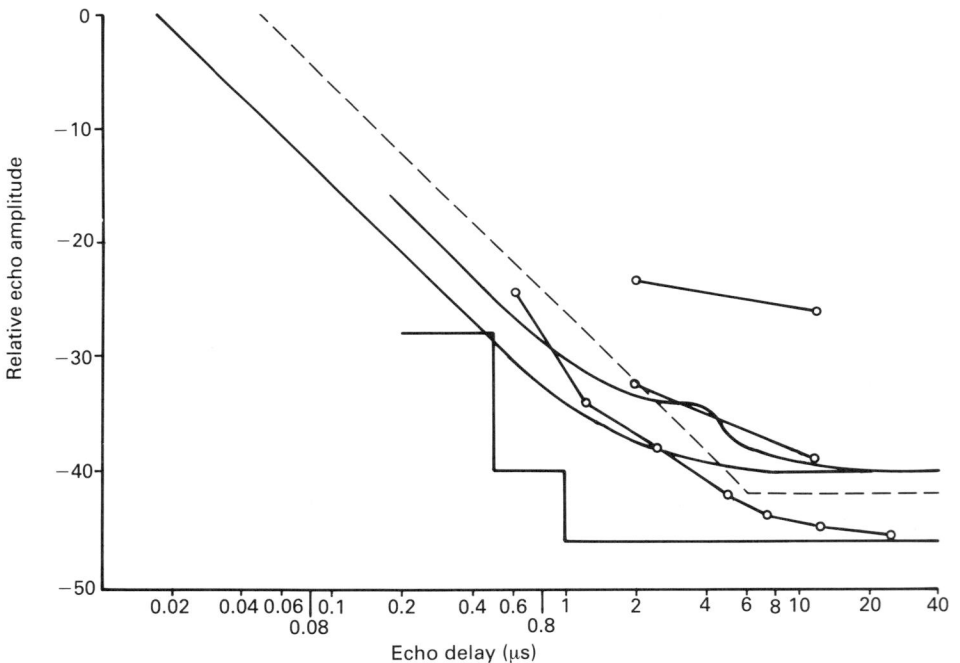

Fig. 7.8 — Ghosting perceptibility curves produced by different researchers to illustrate the various tolerances that they reported on echo amplitude as a function of echo delay (from *J. SMPTE*, May 1953 [1]).

original work, CCIR rec. 654 suggests that the impairment characteristics illustrated in Fig. 7.9, corresponding to the case of an undistorted echo which is delayed by one

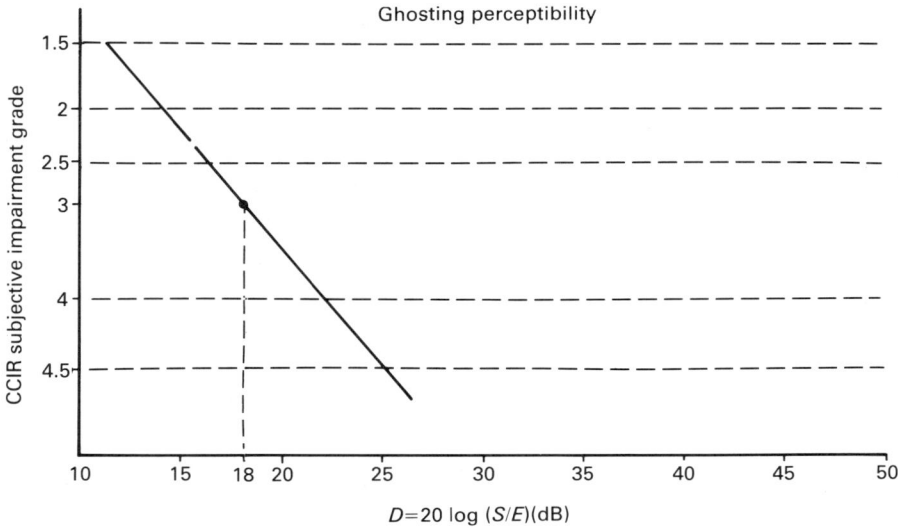

Fig. 7.9 — Impairment characteristic for an undistorted positive echo having $1\,\mu s$ delay. The impairment characteristic D depends upon echo amplitude E and signal amplitude S, so that D = $20 \log (S/E)$ dB (CCIR REC. 567).

microsecond from the main signal, should be used for assessment purposes. The recommendation also suggests that the subjective effect of echoes delayed by time periods other than one microsecond can be obtained by subtracting the correction factor obtained from the curve shown in Fig. 7.10. The impairment values shown relate to those detailed in CCIR rec. 500/2 [4], and the corresponding effects on the picture are generally explained as:

1. Very annoying
2. Annoying
3. Slightly annoying
4. Perceptible but not annoying
5. Imperceptible.

The information given in CCIR rec. 654 represents the combined results of many year's work by different organisations, and can therefore be regarded as authoritative, but the fact that different researchers have, over the years, produced some substantially different curves may be taken as a firm indication that the effect of ghosting depends upon a good many variables, some of which are difficult to control. The subject matter of the picture plays a big part in determining whether or not

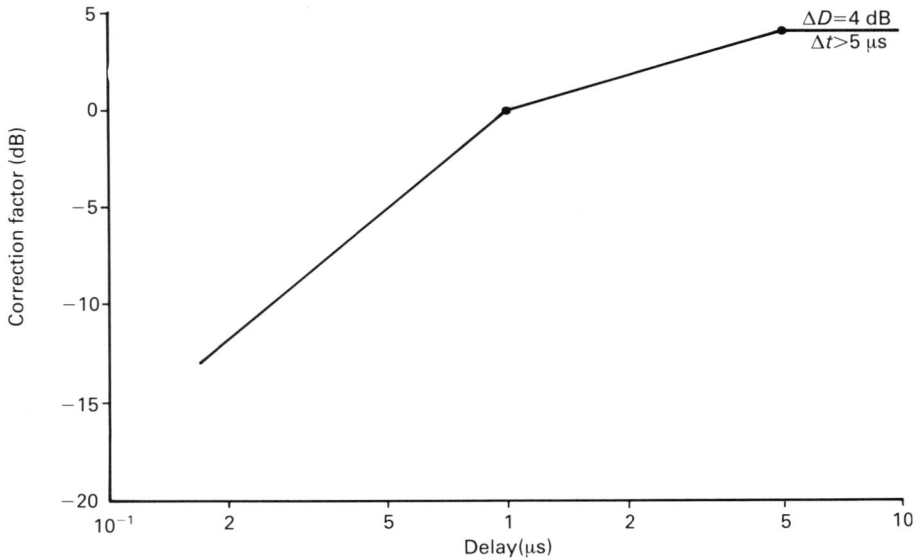

Fig. 7.10 — Correction factor to be subtracted from the value of the signal-to-echo ratio in Fig. 7.9 for use with other values of delay (CCIR REC. 567).

reflections can be seen, and sharp-edged captions often show up problems that are not noticeable on even quite-detailed pictures. The assessment situation is made even more difficult by the fact that the degree of annoyance caused by echoes depends upon whether there is just one, very rare in practice, or whether cumulative effects are seen. Some echoes are also distorted versions of the orginal signals which can have different phase and amplitude characteristics from the original signal, giving rise to combinations of positive and negative echo signals. Although it is difficult to give any meaningful figures for day-to-day assessment of reflections, it may be useful to know as a sort of 'rule of thumb' that it is often suggested that a reflected signal 30 dB down on the main signal and 0.5 μs away from it is just visible.

7.3.1 Echo-rating

As it is difficult to apply regular subjective testing to a practical system, objective test methods have been designed, and full details are given in the cable engineers' 'bible', IEC Standard 728, section 14.3.

Basically the method used takes into account both the time displacement and the amplitude of an echo at any particular point in the system where the measurement is made. Special test pulse signals, described fully in CCIR Recommendation 567, are applied at the input to the cable system, and the same pulse signals are displayed on an oscilloscope and measured at the point along the cable system where the echo-rating is to be assessed. Calibrated graticules drawn on the face plate of the oscilloscope enable the received pulse to be compared with the original, and a measure of the degradation due to reflections can then be made. The test signal is known as a sine-squared pulse with a half-amplitude duration equal to $2T$, where T is the time period appropriate to the TV signal under consideration. As an example, for the 525-line NTSC system, $T = 125$ ns, whereas for the 625-line system, $T = 100$ ns. Fig. 7.11 shows how the measurements are made, so that the echo-rating can be assessed.

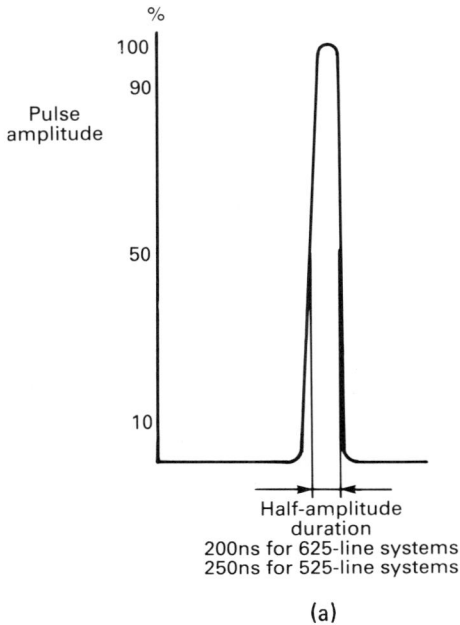

(a)

Fig. 7.11 — (a) Sine-squared $2T$ pulse signal for echo-rating tests. $T = 100$ ns for 625-line signals and 125 ns for 525-line signals. (b) Echo-rating graticule used on oscilloscope (© IEC 1986).

When the pulse-testing method was first developed, many viewers, technical and non-technical, were asked to assess the effect of echoes in terms of their annoyance value, and the tolerance limits were arrived at as a result of these tests. Any echoes or group-delay non-linearity will show up as 'ringing', either before or after the $2T$ pulse, and the amount of this ringing can be measured, and related to the picture impairment that it will cause.

Using the test methods described, the echo-rating as measured at any system outlet is not allowed to exceed 7%, according to IEC Standard 728 Section 35, or 6% according to British Standard 6513:3:3.8.

Although it would be usual for initial tests of this type to be carried out using full-field test signals, one of the beauties of this method is that the pulses can be continuously added to the television picture signal on the blank lines in the vertical blanking period. These so-called Vertical Interval Test Signals — VITS — generally have many other test waveforms added, in addition to the $2T$ pulse, in order to

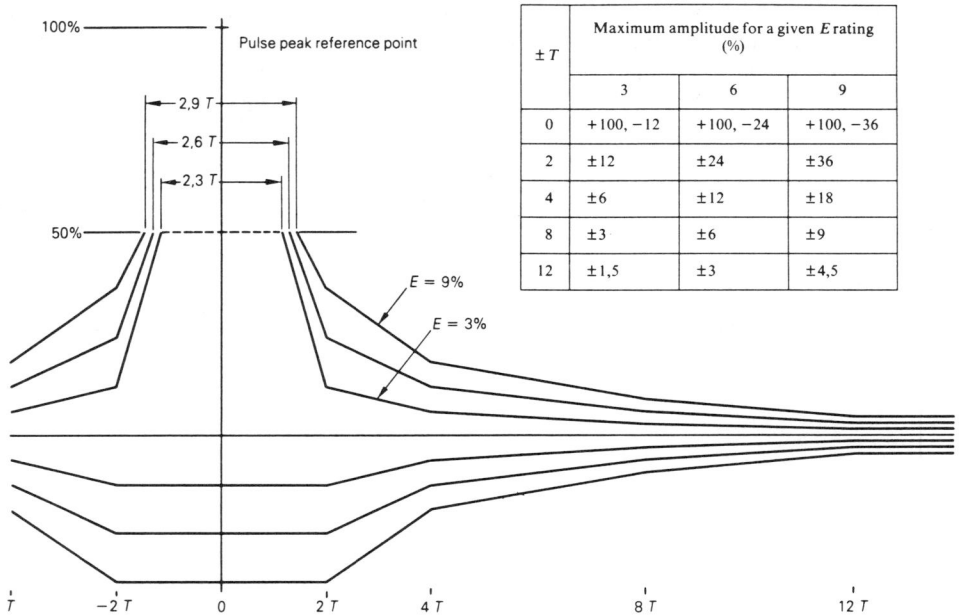

± T	Maximum amplitude for a given E rating (%)		
	3	6	9
0	+100, −12	+100, −24	+100, −36
2	±12	±24	±36
4	±6	±12	±18
8	±3	±6	±9
12	±1,5	±3	±4,5

Fig. 7.11b. (I.E.C. STD. 728)

provide a continuous assessment of the technical performance of the system (Fig. 7.12). The test signals can be obtained from a special generator operated by the wired system operator, or taken from the off-air broadcast channels, an arrangement which is much cheaper, but less flexible, for the operator.

These days it is not even necessary to have a technician with an oscillosope to carry out this type of monitoring, since automatic test equipment is available which continuously makes measurements of the received VITS waveform, compares these measurements with known limits, and alerts the maintenance staff in the case of any problems arising (Fig. 7.13). Such equipment is, however, costly, and only the larger cable systems could justify the financial investment required.

7.4 EXTRA REQUIREMENTS FOR TELETEXT

The teletext data pulses that accompany almost all television transmissions in Europe take the form of specially filtered and shaped NRZ raised-cosine pulses at a data-rate of 6.9375 Mbit/s, and an example is shown in Fig. 7.14.

In order to be sure that pulses at this very fast rate can be decoded accurately, it is necessary to ensure that the echo performance of a system is such as not to degrade the pulse shape. In Chapter 2, while dealing with off-air teletext reception, the concept of using an 'eye' diagram to give a quantitative measurement of the 'goodness' of teletext signals was introduced (Fig. 7.15). In a noise-free system, eye-height can be defined as the ratio of the voltage difference between the lowest '1' pulse and the highest '0' pulse at the optimum sampling instant to the basic teletext amplitude, which is the difference between the continuous '1' and '0' levels.

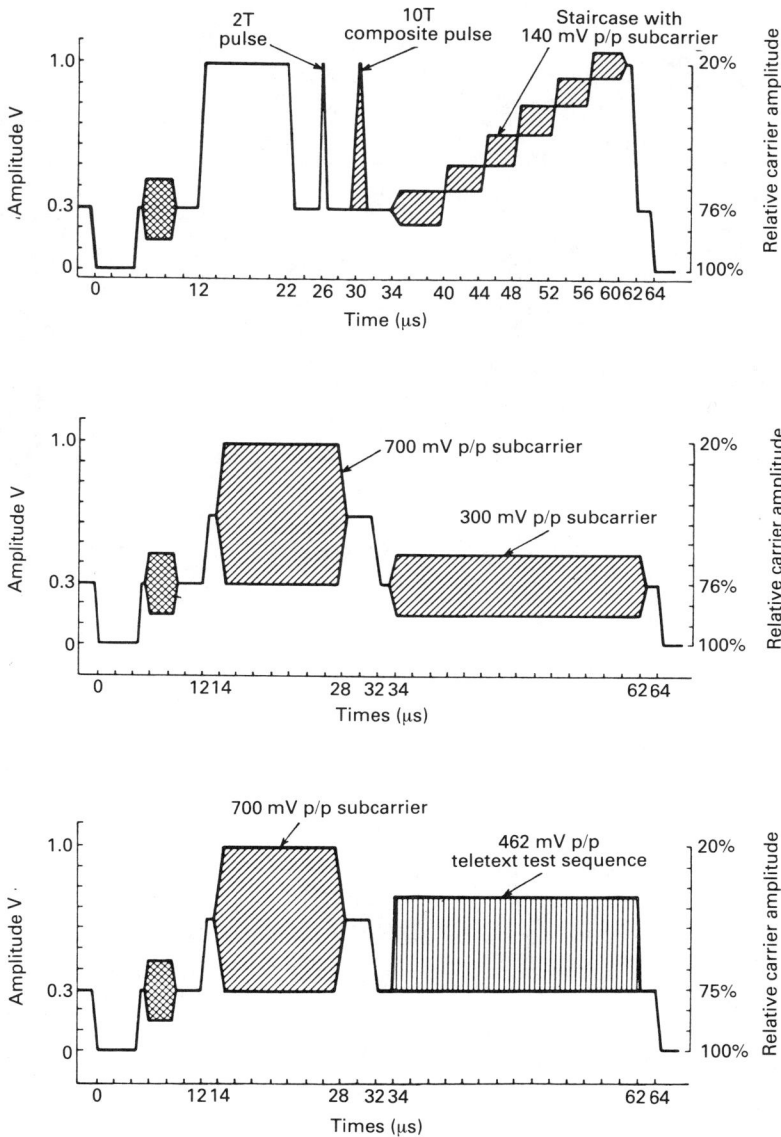

Fig. 7.12 — Typical test signals in the vertical blanking period — VITS. (Courtesy IBA).

Since teletext data pulses are effectively only 144 ns apart, it is not surprising that short-term echoes, which would have no visible deleterious effects on a television picture, can lead to problems for the decoder which is trying to sort out the presence or absence of '0's and '1's in the 6.9375 Mbit/s signal. Many practical measurements have been carried out to try to establish a relationship between the echo performance

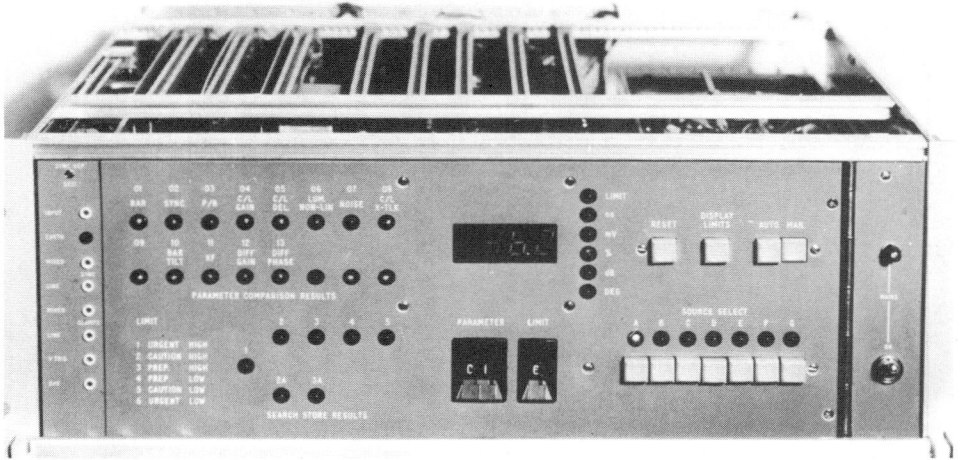

Fig. 7.13 — Automatic VITS measurement equipment (Courtesy IBA).

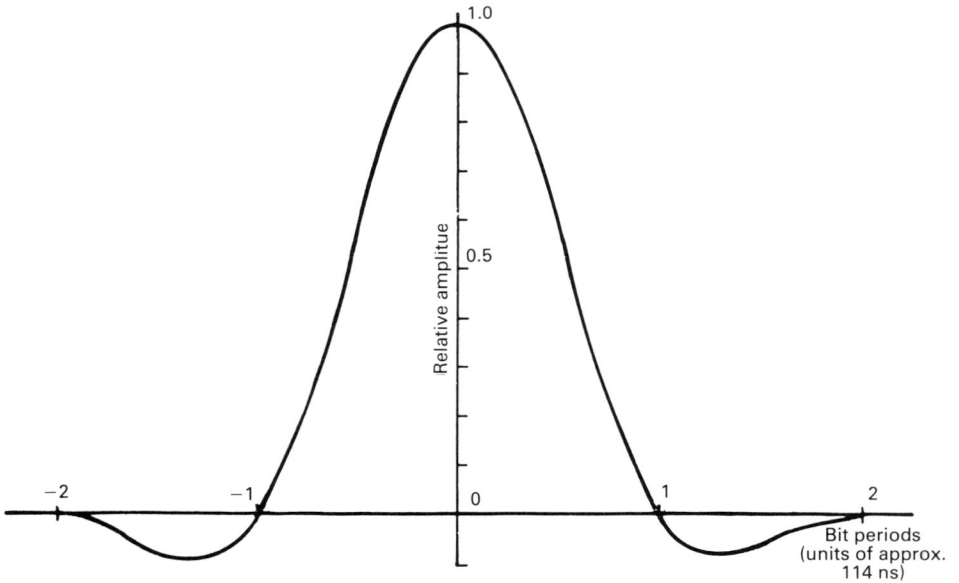

Fig. 7.14 — Typical shape of a teletext data pulse. (UK Broadcast Teletext specification 1976).

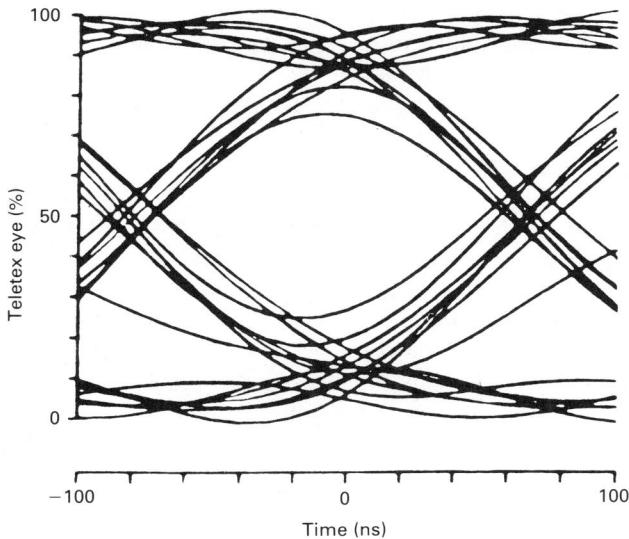

Fig. 7.15 — Eye-height diagram (Courtesy IBA).

wired system, as measured by the conventional $2T$ pulse echo-rating method discussed above, and the eye-height performance of teletext signals, but no significant correlation has been found. It has been found, however, that the slope of the attenuation/frequency curve within the television channel being used has a strong correlation with the measured eye-height, and tests by Italian broadcaster RAI showed that a figure of about 1 dB/MHz gave rise to an eye-height of around 70% when all the other relevant parameters were correct. The use of this slope measurement, in conjunction with the other standard cable system measurements, can be very helpful in cases where the necessary equipment for eye-height measurement is not available. It should be noted that IEC standard 728:34.1 states that the amplitude/frequency response should be such that the variation in gain over a single channel should not be more than ± 2 dB relative to that at the vision carrier frequency, and that the gain shall not vary by more than 0.5 dB within any frequency range of 0.5 MHz.

7.4.1 Teletext performance
Practical tests and surveys of teletext performance on many wired systems have shown that provided the system is well engineered, and that no significant mismatches are present, teletext performance will be perfectly adequate at subscriber outlets. This means that receivers can be presented with teletext signals of better than the 25% eye-height figure that is usually specified as being necessary to provide the bit error rate of 1 in a 1000 that is usually regarded as the acceptable maximum if the viewer is not to notice many decoding errors on the screen. Most VHF and UHF systems can carry teletext signals without problems, and in the early days of teletext, when reports of problems on wired systems were investigated, it was generally found that any problems could be traced to poor installations which had included mis-

matched sections, and that once these mismatches were corrected, the teletext performance became acceptable. Some, but by no means all, of the old HF twisted-pair systems were found to give problems, the eye-height degrading rapidly over long lengths of feeder, and it was also found that severe reductions in eye-height took place in some of the HF-to-UHF converters.

In an ideal world it might be best if the specification for the teletext performance of a wired system could be given in terms of eye-height, since this gives the most precise method of assessing the system performance. Suitable measuring equipment is extremely complex and expensive, however, and practical results have shown it not to be strictly necessary, so the requirements for teletext data signals given in the British Standard 6513:3:3.11 are expressed in a different way. Firstly, the standard television picture echo-rating figure for wired systems of 6% must be achieved. The standard then says that given a suitable test pulse related to the data-rate of the system, negative or positive echoes shall not exceed 10% of the received pulse height. Fig. 7.16 shows what this requirement means in practice.

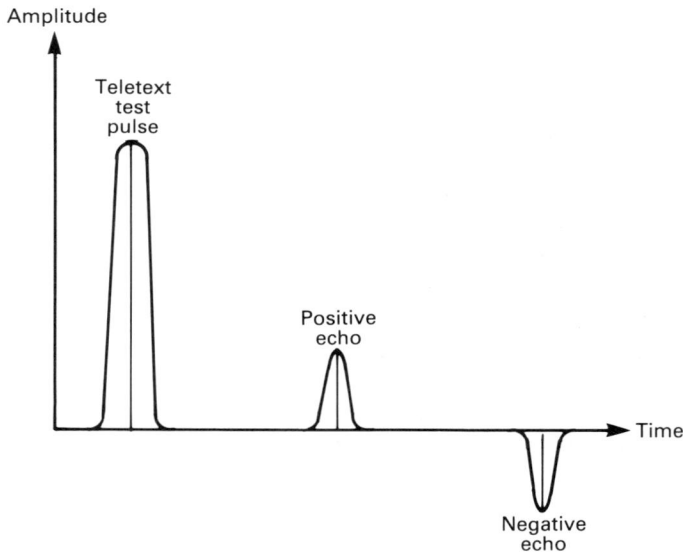

Fig. 7.16 — Echo requirements for teletext. Negative or positive echoes must not exceed 10% of test pulse amplitude.

Equipment for producing accurately calibrated teletext test pulses and test pages, based on the IBA-developed DELPHI unit (Fig. 7.17), is available from Philips, and a professional eye-height measuring set based on the IBA-developed NEMESIS unit is made by Rhode and Schwartz.

7.4.2 Group-delay distortion problems

Distortion which arises from variations in the phase delay of a system over the frequency range required for transmission is known as group-delay distortion, delay

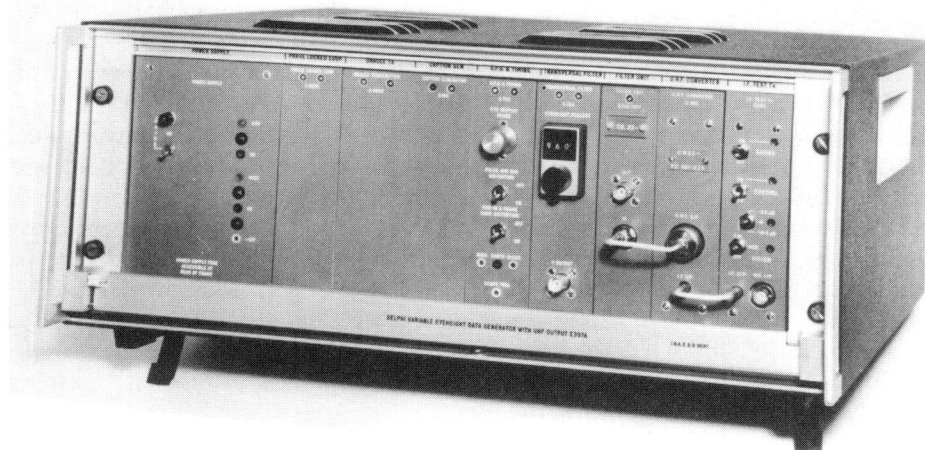

Fig. 7.17 — Teletext test equipment (Courtesy IBA).

distortion or, sometimes, envelope distortion. Practical tests on wired systems showed that the group-delay distortion performance of a system could significantly affect its ability to carry error-free teletext signals, with severe teletext impairments occurring when the group-delay variation exceeded 100 ns, and impairments being visible with variations of as little as 60 ns. British Standard 6513:3:11.1.2 says that the delay inequality between the lowest video frequency and the frequency corresponding to half the data-rate must not exceed 50 ns, and the IEC Standard 728:50.1.1.1 also indicates that the data-delay inequality over the vision channel must not exceed 50 ns.

This can be regarded as a somewhat stringent limit, but in practice, provided that certain precautions are taken at the design stage, problems should not arise. Components such as splitters which contain only passive components and nothing of a highly selective nature will generally have no effect on teletext reception, provided of course that they are properly matched. The main requirement is to ensure that the channel filters and preamplifiers used in the system are not too selective; it has been found that single channel attenuators can ruin teletext reception. Although adjacent channel operation is generally deprecated on wired systems, sometimes these very selective devices are introduced into systems in order to reject an unwanted adjacent channel, and the sharp cut-off characteristics can then affect the signals on the wanted channel, having particularly undesirable effects on teletext reception.

Adjustable-level equalising filters are often used in the input stages of systems using wide-band amplifiers, and it has been found that these introduce both amplitude and group-delay distortions. Usually these filters are multi-stage types designed to reduce the vision carrier at a particular channel relative to the levels of other vision carriers, and the sharp filter characteristics produce group-delay distortions that can significantly impair teletext reception.

As mentioned earlier with regard to HF systems, channel convertors can introduce group-delay distortions, mainly because of the need for a band-pass filter on the input. In such convertors it is important for the local oscillator to remain very stable, since any variations are equivalent to the television receiver being mistuned, and mistuning of receivers can cause teletext reception to be impaired. In 1978, UKIBA experimental work showed that it was possible for receivers to be mistuned without the results becoming apparent on pictures, but resulting in poor teletext reception, and this has since been confirmed by others, so that the importance of ensuring accurate and repeatable tuning of receivers is now well accepted by manufacturers. The figure of ± 75 kHz as the maximum permitted variation from the nominal frequency of a television signal, which is given in IEC 728:35, was arrived at by finding the variation that caused a 'just-noticeable' degradation on pictures, and this tolerance is generally regarded as not tight enough to ensure that teletext signal reception will be optimum. Receiver manufacturers nowadays ensure that the tuning is phase-locked, but some of the frequency-synthesised tuners can still give problems if sufficient thought has not been given to their design.

7.4.3   Performance requirements for other data systems

Since large CATV system operators need to distribute the widest possible range of programme material in order to provide their customers with an attractive service, the use of signals from distribution satellites has become widespread, but since these have so far been 'standard' format signals in NTSC, PAL or SECAM, they have not provided any significant extra problems for the operator, except that he has to cope with various scrambling systems. The more powerful Direct Broadcast Satellites (DBS) that are now available to European viewers have taken advantage of new developments in television to carry a different and significantly improved type of television signal, known as MAC, an abbreviation for Multiplexed Analogue Components. The MAC system is described in Chapter 10 but our interest here is in the data part of the MAC signal, which can be used to carry multiple sound channels or teletext-type data services. Whereas the 6.9375 Mbit/s teletext data pulses that we have considered so far have been using two-level binary coding, the MAC data signals are at a much higher rate, either 20.25 Mbit/s for the so-called D/MAC system, or a sub-set at 10.125 Mbit/s for the D-2 MAC system, and binary data at these rates would require far more bandwidth than the average existing cable system can provide. Multi-level coding is therefore used to reduce the bandwidth requirements, and the particular coding system chosen for the DBS broadcasts is a three-level one, known as duobinary coding. The initial letter of duobinary has given rise to the names D-MAC and D-2 MAC, which are applied to the satellite broadcasting systems, and, as we shall see, to the corresponding signals carried on cabled distribution systems.

7.5   DUOBINARY CODING

Duobinary coding is a specialised form of binary coding in which a binary zero in the input digit stream is represented by a binary zero in the output stream, but where an incoming binary '1' can cause a change in the level of the output pulse stream which depends upon the number of zeros that have occurred since the previous '1' was received. Thus a zero-level of voltage always represents a zero bit in the pulse train, whereas a '1' bit can for example represent a positive voltage level if the number of zeros since the previous '1' bit is even, or a negative voltage level if the number of zeros since the previous '1' bit is odd (Fig. 7.18). Strictly speaking, if the number of '0's since the last '1' bit is even, no change occurs, (i.e. if the previous '1' pulse was positive, the next '1' pulse is also positive). If, however, the number of '0's since the last '1' is odd, a change takes place, so that if the previous '1' pulse was positive, the next '1' pulse would be negative.

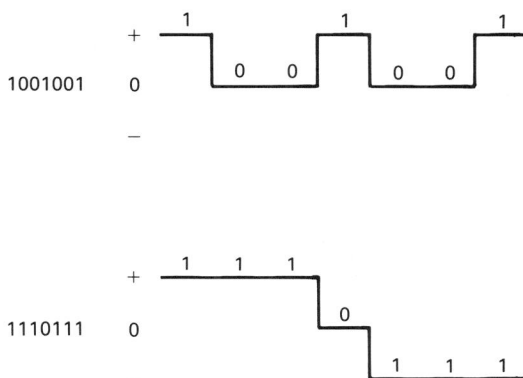

Fig. 7.18 — Binary signals and their duobinary representations.

Duobinary signals are therefore three-level, or ternary, signals, which have a greater data capacity for a given bandwidth than standard binary signals, which is theoretically about twice that of binary, but somewhat less in practice. The signal-to-noise ratio of the three-level signal will be worse than that of the standard binary signal.

To generate duobinary signals, the incoming binary data are filtered, delayed by one bit interval, inverted and then added to the undelayed version of the input signal.

7.6 CARRYING BROADCAST SATELLITE DATA SIGNALS ON CABLED SYSTEMS

DBS signals are broadcast on FM with a bandwidth of about 27 MHz, so it is impracticable to find room for this amount of spectrum space for each channel on a present-day cable system, although one could perhaps conceive of this being possible on future fibre-optic systems. Since all the other television signals are distributed in AM-VSB form on cable systems, it therefore seems sensible to convert the MAC signals to AM-VSB for cable distribution. If the data part of a D-MAC signal at 20.25 Mbit/s is demodulated at the cable system head end, the end result is a duobinary-coded three-level baseband signal at 20.25 Mbit/s. The minimum band-width required to distribute this in the cable system is about 10.125 Mbit/s, plus the width of the vestigial sideband, which gives a total bandwidth of around 11.37 MHz, a significant improvement on the bandwidth of around 15 MHz that would be needed if simple binary coding were to be used.

This is still too much for many of the existing rather-ancient wired systems, which have channel spacings of only about 7 MHz, to cope with, and for these some means of reducing the required bandwidth had to be found. The chosen solution was to transmit the digital sound and data signals at only half the data-rate of D-MAC, which has given rise to the so-called D-2 MAC system, which would have been more appropriately named half-D or D/2 MAC. The 10.125 Mbit/s duobinary data signals can be carried comfortably within a 7 MHz cable channel. Although the reasons for the adoption of D-2 for such cabled systems can be clearly seen, it was somewhat surprising that the French and West German governments chose to use D-2 for their DBS, since the amount of data capacity available was therefore halved, so that only four high-quality sound channels can be transmitted, whereas 20.25 Mbit/s D-MAC allows for eight.

As might be expected, the three-level duobinary coded signals are found to have a worse signal-to-noise ratio performance than an equivalent binary signal, but on practical cable systems where the signal-to-noise ratios are invariably already very good, this proves to be no disadvantage.

The eye-height of duobinary data signals can been measured in a similar manner to that of binary signals, and practical tests have shown that it is possible to reduce the bandwidth from the theoretical 11.37 MHz to about 10.25 MHz without the eye-height degrading very rapidly, whereas with standard binary data the eye-height falls off very quickly as the available bandwidth is reduced.

The application of D-MAC and D-2 MAC signals to cable networks has not yet found its way into the IEC standards publications, but initial work by the IBA [2] has shown that the effects of various system deficiencies are not dissimilar to those when binary data are used. Echoes some 20 dB below the wanted signal at a spacing of 100 ns can cause significant eye-height degradation. Tuning errors of 100 kHz were also found to cause significant eye-height reductions, and group-delay ripples of 100 ns also caused problems (Fig. 7.19).

Comparing these figures with those from the earlier part of this chapter, it will be seen that a system which can carry teletext signals satisfactorily should be capable of carrying D-MAC data signals without any problems. This has been borne out in practice by tests that have been carried out on several existing wired systems, where D-MAC signals have been carried without significant degradation or interference to existing services using the same system.

REFERENCES

[1] P. Mertz, Influence of echoes on television transmission, *J. SMPTE*, May 1953.
[2] H. J. O'Neill and P. A. Avon, The distribution of C-MAC in cable systems, International Broadcasting Convention Proceedings, 1986.
[3] C.C.I.R. Recommendation 567. Transmission performance of TV circuits.
[4] C.C.I.R. Recommendation 500/3. Method for subjective assessment of quality of TV pictures.

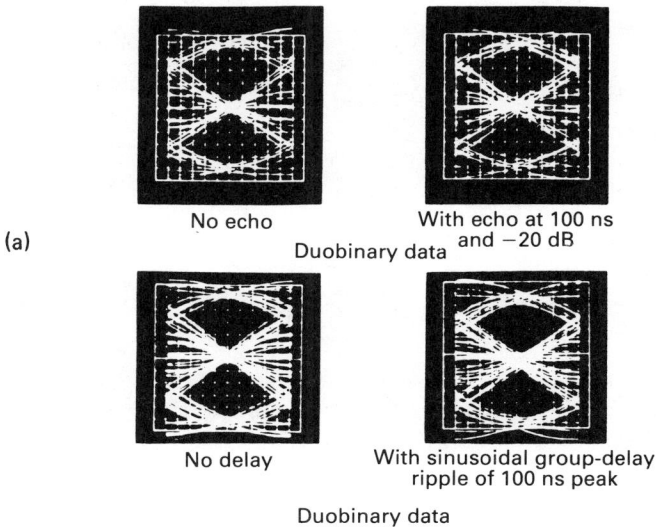

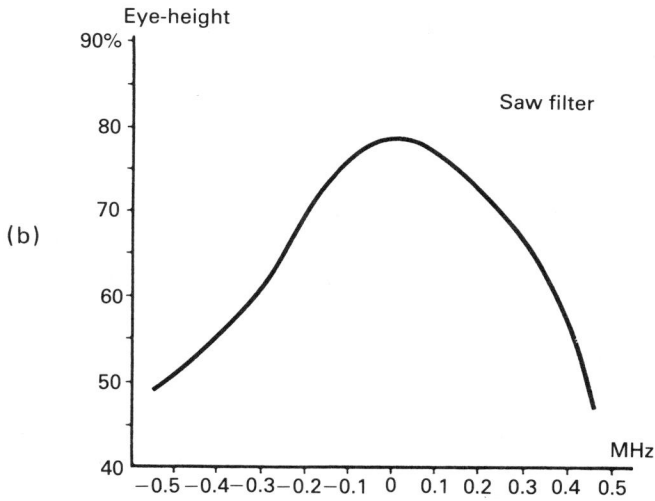

Fig. 7.19 — Theoretical effects of delay and echoes (a), and of mistuning (b) on duobinary data
(Courtesy IBA).

8

Fibre-optic feasibility

As a result of many reports in both the popular and the technical press, it is commonly believed that modern cable systems must use fibre-optic cables, but although these technological light-pipes have many advantages over conventional cables, they are not necessarily the answer to every cable-man's prayer at the present time, and we have seen elsewhere in this book that systems using coaxial cable can provide a wide range of services and facilities.

Although the costs of fibre-optic and coaxial cables are often compared, and metre for metre some coaxial and fibre prices are very similar, for a true comparison it is also necessary to take into account the costs of the modulators and demodulators that are needed for the optoelectronic interfaces to convert the television and radio signals to and from the infra-red light signals that pass along the glass fibres. The light source is usually either a high-performance light-emitting diode or, preferably, a laser which can be switched on and off at high speed to send a high-powered burst of light into the fibre. The receiver usually makes use of some form of photo-diode. Such total cost comparisons are often unfavourable to fibre-optics at present-day prices, and since it is perfectly possible for a well-engineered coaxial system to provide all the service that we can currently envisage, it is not surprising to find that many of the currently planned cable schemes will be using the well-tried coaxial cable systems whose mature technology is well understood. Even so, many of the new systems are being planned as switched-star layouts which will be able to fairly readily make the change to fibre-optic technology at some time in the future when the operator comes to believe that this type of system can be installed and operated economically. Even the most diehard of coaxial cable supporters realises, however, that the in-built advantages that glass-fibre-optic systems possess must mean that sooner or later this new technology will prevail. We shall therefore look at various aspects of the use of fibre-optics in cabled distribution systems in order to understand the reasons why they will be adopted, and the advantages and disadvantages that will accrue.

8.1 FIBRE-OPTICS VERSUS COAXIAL — THE PROS AND CONS

Optical fibres

Low attenuation; wide repeater spacing.

Higher bandwidth than coaxial cables; greater transmission capacity.

Immune to electromagnetic interference; do not radiate.

Intermodulation unlikely.

Physically small, leading to easier installation.

Lightweight; weight of wiring looms in vehicles and aircraft much reduced.

Secure; difficult to tap or detect cables.

Inherently safe; no risk of sparking or overheating in danger areas.

Non-conducting; no earth loop problems; provide electrical isolation.

Non-conducting; cannot carry electrical power for repeaters, etc.

Input/output equipment currently expensive.

Jointing difficult; comparatively expensive.

New technology; still much to learn.
Long-term fibre reliability unknown.

8.1.1 Attenuation at different wavelengths

It is important to note straightway that the 'light' signals which pass along the glass fibres are not in fact visible light, which uses wavelengths between about 455 and 750 nm, since these would be severely attenuated in present-day cables. Instead, lower-frequency signals with wavelengths of about 850–1500 nm, just beyond the limit of visibility in the infra-red range, are used.

Any glass-fibre cable exhibits losses as signals pass along its length, and these are usually expressed in dB/km. Fig. 8.1 shows how the losses of a typical fibre might vary

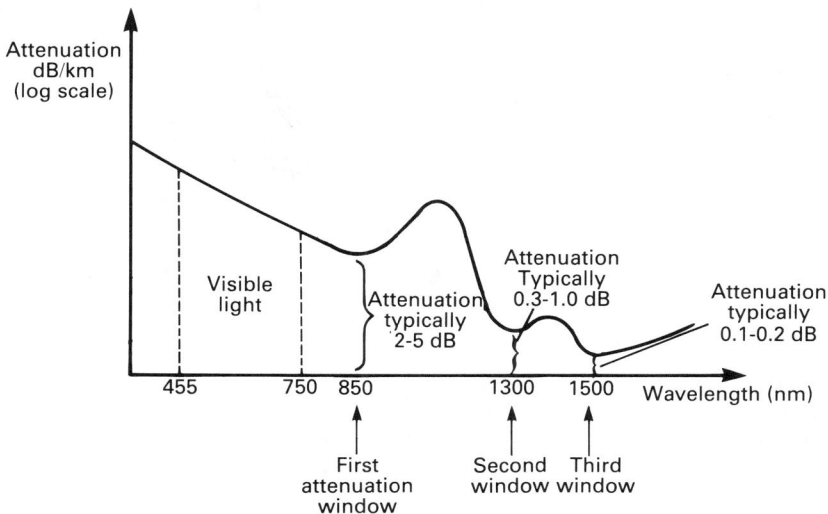

Fig. 8.1 — Fibre losses as a function of wavelength.

with the wavelength, and it will be seen that there are three regions where the attenuation is at a minimum. These wavelengths where the lowest losses occur are known as 'windows'.

It will be seen from the graph that the first window occurs at wavelengths of about 850 nm, and a typical fibre might well exhibit attenuations of perhaps 2–5 dB/km in this region. The second window occurs at about 1300 nm, where losses of 0.3–1 dB/ km would be typical, and a third window is found at about 1500 nm where the lowest losses, perhaps 0.1–0.2 dB/km, occur. In dealing with these 'light' waves it is usual to deal in terms of wavelength, rather than frequency, but consideration of the above figures shows that the greater attentuations occur at the shortest wavelengths, ie. at the highest frequencies, and this is just what we would expect with a standard coaxial system using VHF or UHF frequencies. At the present time there are problems in providing solid-state light sources at an economical price for the 'third-window' frequencies, so that most systems are designed to make use of the first and second windows, with the 1500 nm signals being reserved for only the longest of telecommunications links, although this may well change as component developments take place.

8.1.2 Types of fibre construction

Just as there are many different types of coaxial cable, which have been developed over the years to satisfy many different requirements, so there are different kinds of fibre-optic cable, and it is important to be able to distinguish between these in order to understand some of the arguments that are being put forward by both the pro- and the anti-fibre lobbies. In the research laboratories of the world's telecommunications specialists there are fibre-optic cable systems which are capable of carrying a couple of television channels in digital form over a distance of greater than 100 km without any amplification, but although it is these systems that make the headlines, the more typical cables which are actually being produced at the moment and which are available for purchase can only operate satisfactorily carrying from four to six television channels over distances of up to about 10 km without regeneration or amplification. This may at first seem disappointing, until you remember that conventional coaxial-cable-based systems require amplification at least every kilometre, and sometimes it is more realistic to amplify the signals every 250 m.

Typical optical fibres are between about 30 and 600 μm in diameter, and there are three common forms of basic construction, but before we look at these in detail it may be useful to remind ourselves of the basic physical laws of optics which make it possible to guide light along a fine glass fibre.

8.2 PRINCIPLES OF OPTICAL TRANSMISSION

Light travels more slowly in glass than in air; this is because glass has a greater density, and the term 'refractive index' is used to relate the speed of light in a vacuum to the speed of light in glass.

$$\text{The refractive index of any substance} = \frac{\text{Speed of light in a vacuum}}{\text{Speed of light in the substance}}$$

If a ray of light carried in a piece of glass approaches the glass/air boundary, its behaviour at that boundary will depend upon the angle at which the light approaches it, and Fig. 8.2 shows the effects at different incident angles.

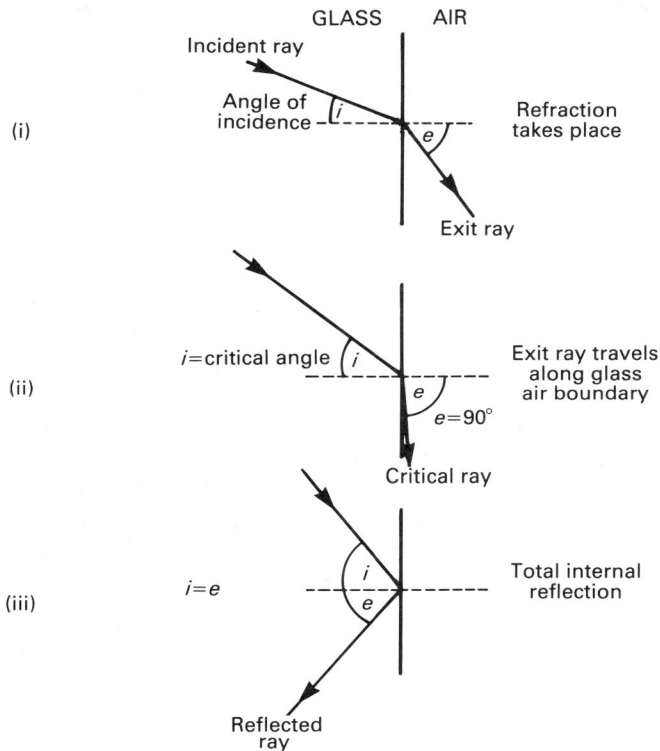

Fig. 8.2 — Behaviour of light at a glass/air boundary.

Snell's law of reflection showed that the angle between the incident ray and the normal to the boundary (i) could be related to the angle between the exit ray and the normal (e) by the formula:

$$\frac{\sin i}{\sin e} = \frac{\text{Refractive index of glass}}{\text{Refractive index of air}}$$

This helps us to understand that at modest angles of incidence the exit ray will behave as shown in Fig. 8.2(i). As the angle of incidence increases, there will come a 'critical angle' at which the ray of light emerges along the glass/air boundary (Fig. 8.2(ii)), and if the incident angle is increased further than this critical angle, virtually all the light will be reflected back into the glass, a situation known as 'total internal reflection' (Fig. 8.2(iii)).

Fig. 8.3 shows, in a much simplified form, how total internal reflection can be used to trap light rays in a narrow cylinder of glass by a whole series of such reflections, and it is on this basis that all our fibre-optic cables operate.

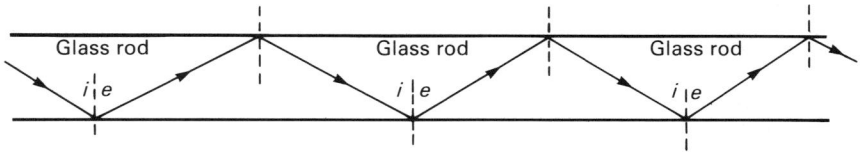

Fig. 8.3 — A light ray trapped in a glass cylinder by total internal reflection.

In real life, things are not quite so simple, of course, and we find that if a simple glass cylinder is used, scratches and imperfections on the surface and contamination by grease and dirt can cause the light to escape. If grease does contaminate the glass surface, it has an effective refractive index much closer to that of glass than of air, so the light will pass out through the surface of the glass into the grease, and will therefore be lost as far as travelling down the fibre is concerned. This problem is overcome by surrounding the simple glass cylinder with a sheath or cladding. The infra-red light signals are then carried by the central core of the fibre, which is surrounded by a cladding of glass with a slightly lower refractive index than that of the core. Such a construction allows the light to be guided along the core, travelling by total internal reflection along a general zig-zag path, the differing refractive indices of the core and cladding ensuring that light does not escape.

8.3   PRACTICAL FIBRE DESIGNS

The structure of each of three major different types of fibre-optic is considered below.

The step-index fibre (Fig. 8.4) has a sharply defined change in refractive index between the core and the cladding. It will be seen from the diagram that it is possible for the light signals to take various different paths along the fibre, and any practical cable of this type will allow transmission via a multitude of different paths. These paths will vary considerably in length, the shortest path being taken by those rays that start off parallel to the axis of the fibre, and the longest will be taken by those rays that enter the cable at a considerable angle to the axis. Each of these paths, at a different angle, is known as a transmission mode, and this type of fibre is therefore also known as a multi-mode design.

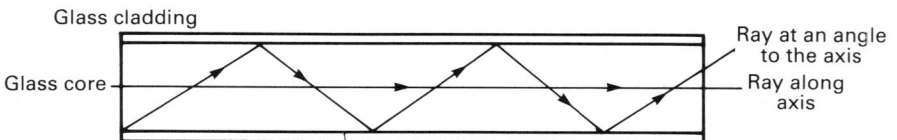

Fig. 8.4 — Structure of a step-index optical fibres.

8.3.1 Dispersion

Since the rays travelling via different modes are obviously travelling over considerably different distances, and since the speed of light will be the same throughout the glass pathway, since the refractive index of the glass is a constant, some rays of light will arrive at the far end at different times from others. This means that if the input signal was a very short pulse of light, this pulse will be subject to many varied transmission delays as it passes along the fibre, and the short input signal will therefore arrive at the output as a signal which is dispersed over a considerable period of time. This phenomenon is known as dispersion (Fig. 8.5). and this restricts the rate at which data may be transmitted, since if many narrow pulses are transmitted during a short interval, the pulses travelling via the different modes will become intermixed, and it may not be possible to sort out one pulse from the other at the receiving end. Since the pulses effectively spread out in time, the amplitude of each pulse will necessarily be reduced, and this phenomenon is called intermodal dispersion.

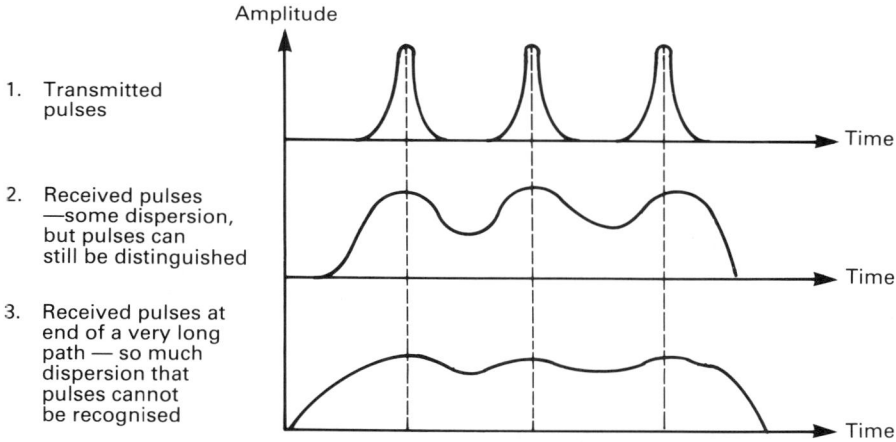

Amplitude

1. Transmitted pulses

Time

2. Received pulses —some dispersion, but pulses can still be distinguished

Time

3. Received pulses at end of a very long path — so much dispersion that pulses cannot be recognised

Time

Fig. 8.5 — The effects of dispersion.

The amount of dispersion introduced by a length of fibre is generally measured by comparing the widths of the input and output pulses, the measurements usually being given in nanoseconds. The test pulses used are of a Gaussian shape, and the pulse-width measurements are taken at the half-/amplitude points (Fig. 8.6).

The dispersion of practical fibres is frequently quoted in terms of ns/km, but although this implies a linear relationship, which gives fairly accurate answers over short lengths of fibre, it has been found that when very long fibre paths are used, the dispersion is proportional to the square root of the fibre length. This reduction in the dispersion that might have been expected comes about because it is found that even

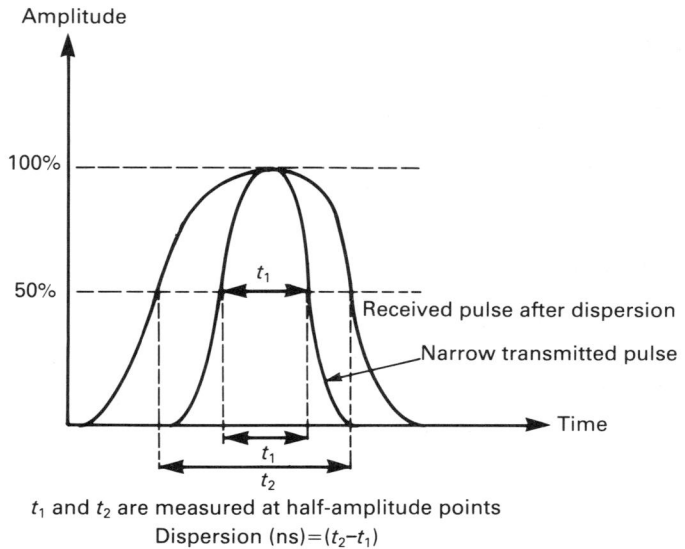

t_1 and t_2 are measured at half-amplitude points
Dispersion (ns)=(t_2-t_1)

Fig. 8.6 — Dispersion measurement.

the 'on-axis' rays suffer from changes in the ray angle as they pass through long lengths of fibre, and are therefore delayed fractionally more than might have been predicted.

An alternative method of taking account of the effects of dispersion is to quote the 'fibre bandwidth'. This is proportional to the reciprocal of the dispersion, and is usually specified as the frequency at which the amplitude of the modulating signal is reduced by 3 dB.

8.3.2 Graded-index fibres

Another type of fibre is constructed so that the refractive index does not suddenly change from that of the cladding to that of the core, but is subject to a gentle transition across the diameter of the cable; such fibres are known as 'graded-index' cables (Fig. 8.7). There will obviously be many different paths, over which the light can travel along such a fibre, and it is fairly typical to find that several hundred different paths can be travelled by light following one of these multi-mode fibres. As in the case of the step-index design, some of these paths will obviously be longer than others and it might therefore seem that the light travelling along the longer paths would again take longer to get to the end of the cable than that taking the shorter paths. In fact this does not happen, as a graded-index fibre has a higher refractive index at its centre than at its cladding, and the index varies gradually across the diameter of the cable.

It will be remembered that as the refractive index of the glass increases, so the speed of the light through it decreases, and this means that the light travels more slowly in the centre of the fibre than it does at the edges. The net result of this is that all the various light rays take approximately the same time to travel along a fibre of

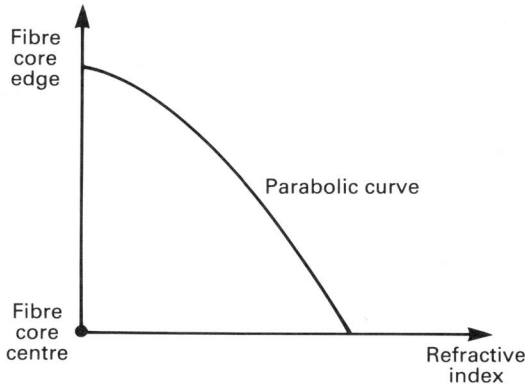

Fig. 8.7 — Refractive-index profile of graded-index fibre.

this type. In order to achieve this effect, the refractive index of the fibre must vary in an almost parabolic manner across the fibre, as shown in Fig. 8.8.

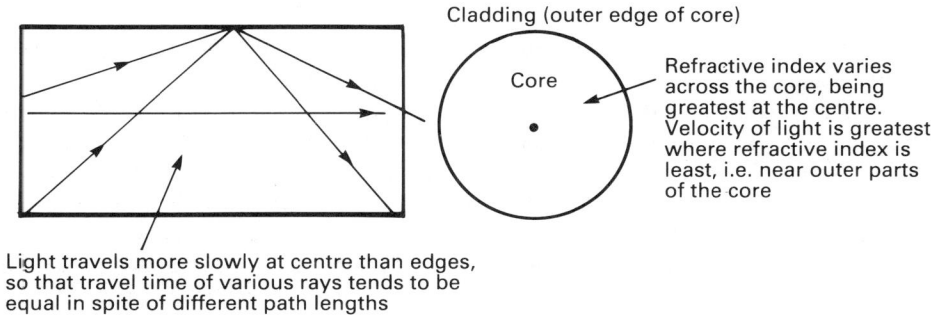

Fig. 8.8 — Graded-index fibre.

This is important because, as we have noted already, if there were appreciable differences in the propagation time, the short pulses of light that make up the wanted signal could effectively spread out in time, causing great confusion, as the receiving end would never be sure when one pulse had ended and the next one had begun. It is not easy to ensure that the refractive-index profile of this type of fibre matches the theoretical profile exactly, so graded-index fibres will always show some degree of pulse-spreading in spite of the care that is taken in cable construction, but dispersion is reduced by a factor of 10 or more, which provides a useful performance improvement over step-index fibres.

8.3.3 Monomode fibres

Although the graded-index fibre provides reasonable results at a reasonable price there will be definite advantages in the future in using a different form of fibre, the so-called monomode type (Fig. 8.9). The core of a monomode fibre is very much smaller

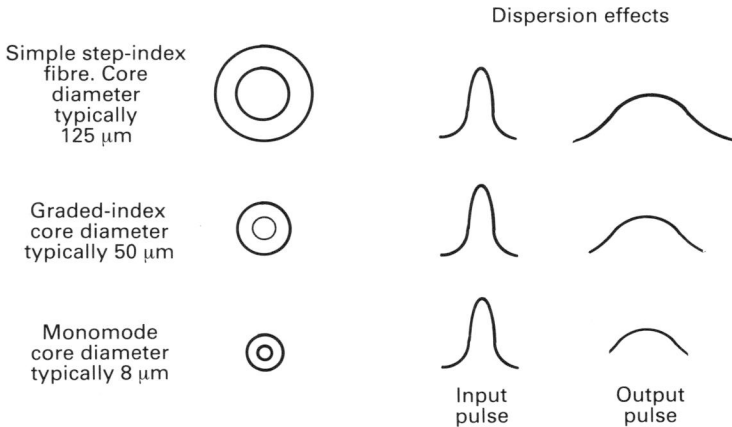

Fig. 8.9 — Comparison of monomide fibre with multimode types.

than that of the multimode type, being perhaps only 5 μm in diameter. Because it is so fine, this type of fibre can only support one single light path for any particular wavelength of light, so that problems with pulse spreading due to different path lengths cannot occur. This means that signals with very large bit-rates or bandwidths can be transmitted over very long distances. It is, however, found that the refractive index of such fibres varies with the wavelength of the light that is being used for transmission, so that the transmission velocity of signals at different wavelengths is different. This can cause another form of dispersion called *intramodal dispersion,* which can limit the (extremely wide) bandwidths that are possible. In order to make the very best use of monomide fibres, therefore, it is necessary to have a light source with a very narrow spectral bandwidth, and the laser satisfies this requirement very well.

Persuading the light signal to enter a fibre is obviously more difficult as the fibre becomes smaller in diameter, and it is important to realise that only those rays that enter the fibre at shallow angles to the axis will satisfy the requirements to be totally internally reflected and therefore travel down the fibre. The acceptance angle shown in Fig. 8.10 depends upon the refractive indices of the air, the cladding and the core, and the sine of this angle, known as the *numerical aperture,* provides a useful measure of the light-gathering ability of the fibre. The light accepted by the fibre will be proportional to the square of the numerical aperture, and to the square of the diameter of the fibre. As long as the diameter of the core is greater than about ten times the wavelengths of the light being used, multimode transmission can take place, but below a critical diameter, usually about 5 μm, monomode transmission

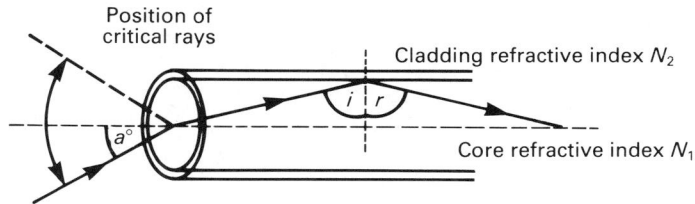

Fig. 8.10 — Acceptance angle and numerical aperture.

can be achieved, and it can therefore be seen that the light source for a monomode transmission must be extremely small.

It is the monomode fibre that makes the headlines these days, and the longest commercial monomode link currently installed carries signals at 140 Mbit/s between Luton and Milton Keynes without any intermediate booster amplifiers. In 1988, British Telecom expect to introduce a 134 km fibre-optic cable between the Channel Island of Guernsey and the south coast of England, again without any repeaters being necessary. Monomode fibre was initially very costly to manufacture, and connectors too are expensive because the machining tolerances needed to obtain minimum-loss connections are extremely tight. At the present time, connectors for all types of optical fibre are more expensive than their coaxial equivalents, but much work is being put into the development of jointing fittings, and it is expected that eventually it will be feasible to join all types of fibre-optic with low-loss connections that will be relatively inexpensive.

Since the benefits of using monomode fibres have been recognised, however, manufacturers have put a great deal of research effort into ways of fabricating usable fibres of this type, and this has led to significant cost reductions, which have resulted in monomode fibres become available at very competitive prices, and it seems that there are now no real reasons for not using this type of fibre in new cabled distribution systems.

8.3.4 Fibre joints
Virtually all methods of joining fibres involve some means of making a butt joint between the ends of the fibres that have been previously prepared in oder to be as flat as possible (Fig. 8.11). A butt joint will suffer from some slight unavoidable reflections where the two surfaces meet, and losses will also occur if the fibres are not perfectly aligned so that all the light entering the joint passes through to the other side. Axial alignment can be difficult, but is vitally important, since if a fibre has a diameter of only 50 μm, the slightest misalignment will cause a large loss.

If the gap between the fibres is filled with an index-matching liquid, i.e. a liquid

Simple butt joint showing how reflections take place

Liquid with refractive index same as that of core

Showing how index-matching liquid
reduces reflection at joint surface

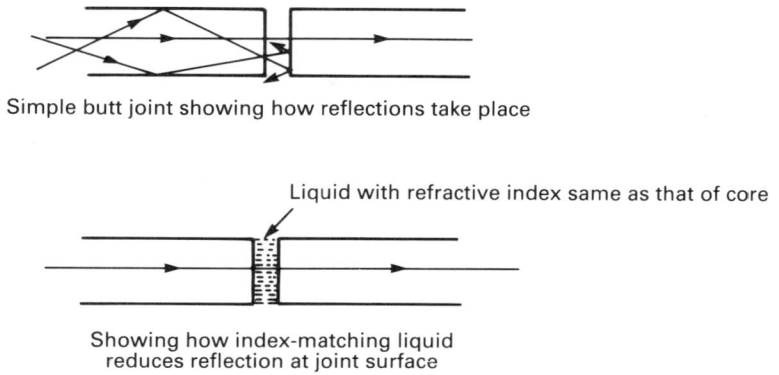

Fig. 8.11 — Butt-joint reflections, and the use of index-matching liquid.

whose refractive index is the same as that of the fibre core, then losses can be very much reduced, and splice losses of about 1 dB or better can be obtained. This method was not thought to be very practicable for joints that may have to be disassembled in the field, but so-called 'elastomeric' joints using this principle are now available.

When a fibre is scribed around with a diamond cutter, it will break cleanly with a good flat end, and simple hand tools are available to carry out this 'cleaving' process. Many forms of mechanical connector can then be used to hold the two ends of the fibre together, the simplest ones usually introducing losses of as much as 2 or 3 dB. Precision ferrules are used to provide some improvement over the simplest joints, and sometimes the fibres are glued into the ferrules with epoxy resin, and the ends ground flat and polished (Fig. 8.12).

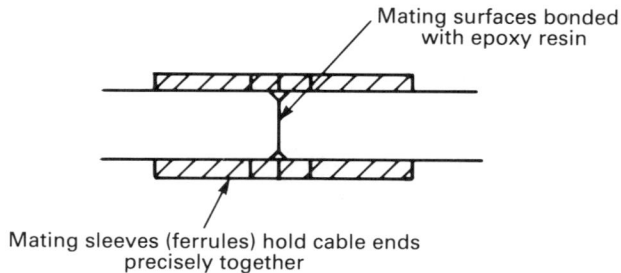

Mating surfaces bonded
with epoxy resin

Mating sleeves (ferrules) hold cable ends
precisely together

Fig. 8.12 — Ferrule connector.

The method of jointing glass fibres that leads to the most efficient, i.e. the least-lossy, joint is known as arc fusion, and this process can now be carried out quickly in the field by means of a battery-powered fusion splicer. This unit can inject light into one side of the joint and then the two halves of the cable are manipulated until the maximum amount of light comes out of the other side, thus ensuring that the two halves of the fibre are optimally aligned. An electric arc is then generated to fuse the two halves together, and losses of as low as 0.1 dB can be achieved with this type of joint.

In general, the larger the diameter of the fibre, the easier it is to achieve a satisfactory join, which is why some operators have so far steered clear of using monomode fibres. For large-diameter fibres in difficult environments where axial misalignment might occur owing to vibration or problems might arise owing to dust particles, an expanded beam connector can be used. As Fig. 8.13 shows, sapphire

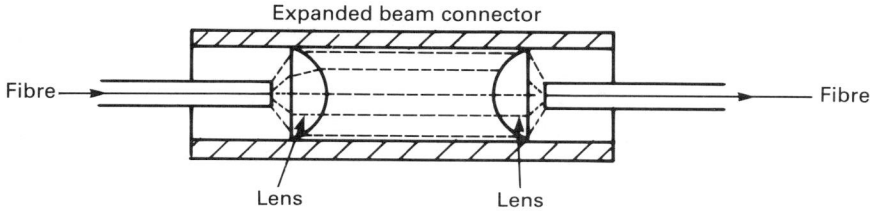

Fig. 8.13 — Expanded beam connector.

lenses are attached to each end of the fibre, in order to provide a large-diameter light beam between the lenses. This beam is relatively insensitive to axial misalignment, although angular misalignment can still cause problems.

8.3.5 Optical taps
Although it is true that only a few years ago it was considered to be impossible to tap signals from optical cables, optical taps are now readily available and form an integral part of several cable companies' plans for distribution networks. Tapping into a fibre-optic cable does, however, require sophisticated splicing equipment and trained users, and so is not yet so convenient as inserting a tap in a coaxial cable. Led by the demand from computer users with requirements to transmit vast quantities of digital data, manufacturers are now making a range of other fibre-optic devices, such as multiplexers, splitters and line-extenders, and it is only a matter of time before these are set to work in CATV systems (Fig. 8.14).

It may be that the most economical way of using fibre-optics in a cable network is to use monomode fibres for the long-haul trunk routes where minimum attenuation is vital, changing to multimode fibres for local connections. There is some disagreement over this, and as to the most sensible economic method of providing the 'drops' or connections to individual subscribers; some people believe that multimode cable will prove the best solution, whereas others think that coaxial connections will be good enough. Both the recently installed Westminster (London) cable systems and the Danish DOCAT system have chosen to use coaxial cables for the final subscriber connection, in spite of using the latest fibre-optic technology for other parts of their networks. In complete contrast, the British Cabletime company uses coaxial cables for its trunk network, followed by low-cost step-index fibre connections from the switching points into the customers' homes.

In a few years' time it may even prove possible to use plastic optical cables for these short runs where attenuation is not too important. These so-called polymer fibres are being developed in America and Japan, and they should be many times cheaper than today's glass fibres. At the moment it is not possible to make such fibres with attenuations below about 150 dB/km, but if this can be reduced to perhaps 35–40 dB/km the plastic cable might well become the ideal distribution medium for short runs.

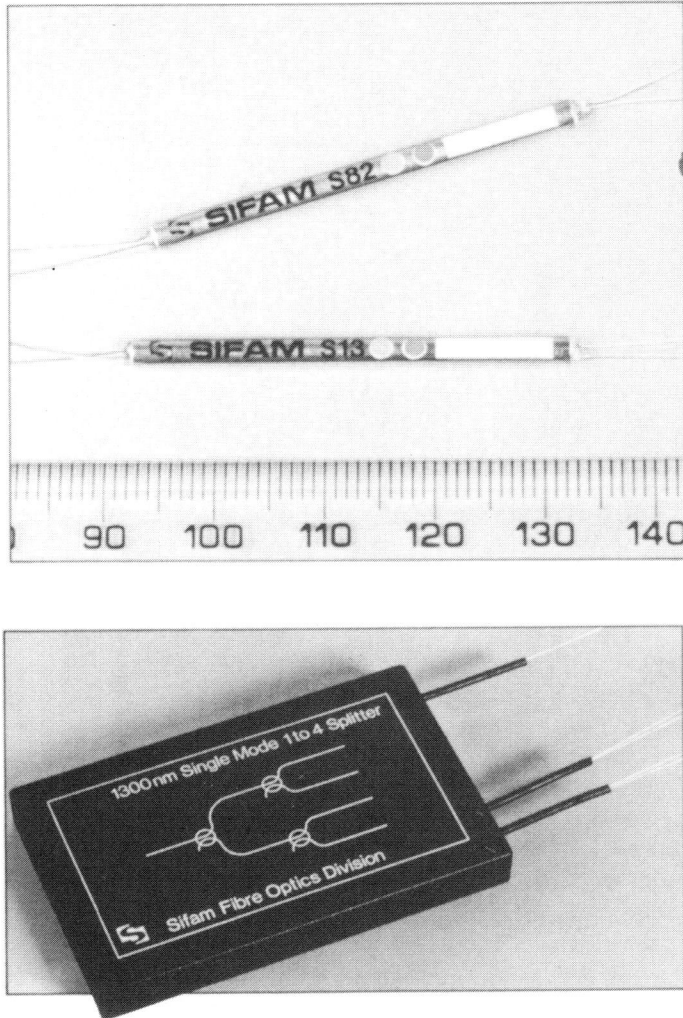

Fig. 8.14 — Photograph of optical couplers and splitters (Courtesy SIFAM).

An interesting possibility that the use of plastic cables could bring is the ability to physically tap off some of the light which travels along the fibre. The AB Electronics company has developed a tap which contains an internally reflecting prism made of sapphire, or a hard plastic. As Fig. 8.15 shows, to tap off signals from the main fibre,

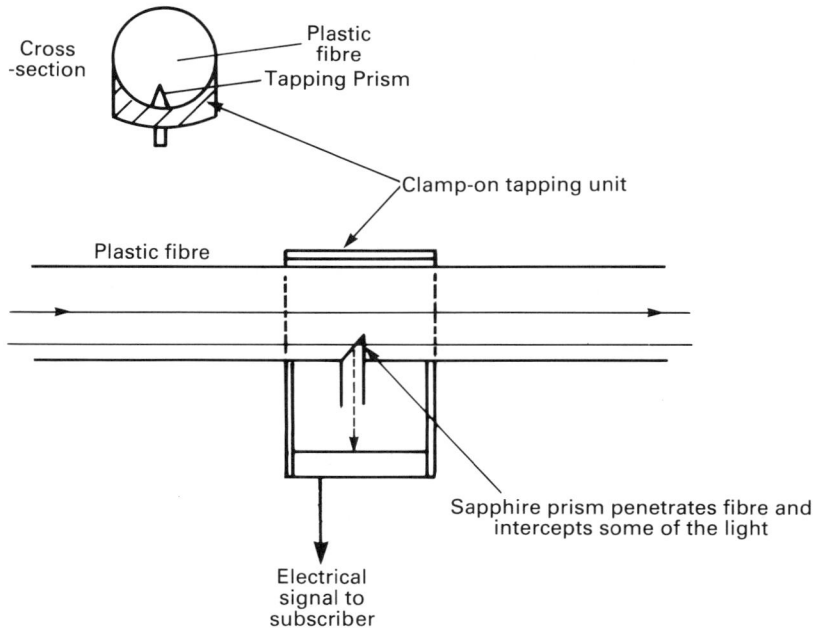

Fig. 8.15 — Clamp-on tap for plastic fibre-optic cables.

this prism is pressed into the fibre core through its cladding, merely by clamping the unit onto the fibre. Some of the light from the main fibre will then enter the prism and be refracted into a transducer, converting it into an electrical signal, but because the tapping prism is itself transparent, very little attenuation is caused to the original signal travelling along the main fibre. So far this device is not suitable for television systems, being more intended for use in localised data-carrying networks, but the idea seems a sound one, and could really take off if suitable plastic cables become available. Similar techniques cannot be used with glass fibres, because the insertion of the prism would break the glass.

The Plessey company has recently discovered that if a fibre optic cable is bent fairly sharply around a curved object, some of the light travelling along the cable will escape through the surface of the fibre, due to scattering. The light which escapes can then be detected by an optical receiver, effectively forming a non-invasive method of tapping signals form the cable.

Optical cables have other advantages as well as their low attenuation characteristics. Since the frequency of the light signals is extremely high the available bandwidth is much greater than for coaxial, which means that optical fibres have a greater potential for the future in applications such as data transfer and interactive services, although it is probably true to say that a properly engineered coaxial system can cope with most of the services which are currently envisaged. The wide bandwidths make signal-equalisation equipment generally unnecessary, so that much of the equipment used in conventional CATV systems can be done away with, with a consequent increase in reliability and reduction in maintenance effort.

8.4   WAVELENGTH MULTIPLEXING

Although the use of extremely-high-frequency light signals, approaching 10^{15} Hz, itself implies that fibre-optic systems can carry enormously high bandwidth signals, various forms of multiplexing can be used to more fully use the capabilities of a fibre. Electrical multiplexing can be used in time-division (TDM) or frequency-division (FDM) formats, but recent developments in optical technology show that a technique known as 'wavelength multiplexing' can be used to increase system flexibility and the transmission capacity of an optic-fibre system.

The basic idea is to carry several different colours of light, i.e. light rays of different wavelengths, along the fibre at the same time. Selective narrow-band optical filtering is used to separate the different colours (wavelengths) at the output. The Marconi multiplexer/demultiplexer unit illustrated in Fig. 8.16 shows how three separate signals in the 'first window' around 850 nm can be used to carry three completely separate signals, with figures being given for the isolation between ports when the device is being used in either role.

Plessey Research have developed a WDM system which allows up to 40 separate channels to be combined along one fibre using a highly selective colour filtering technique known as 'spectrum slicing'.

Wavelength-division multiplexing can also be used with different operating 'windows', and systems using both the 850 and the 1300 nm regions have been built. Much interest is being shown in this technique for two-way interactive services.

8.5   OPTOELECTRONIC INTERFACES — TRANSMITTERS AND RECEIVERS

Since television signals are invariably in the form of electrical signals at the input to a distribution system, it is necessary to convert these electrical signals into optical signals so that they will travel through the optic-fibre distribution system, and then to reverse the process in order to give the receiver the electrical signals it requires.

8.5.1   LEDs and LDs

The transmitters used in all fibre-optic cable television distribution systems use either a light-emitting diode (LED) or a laser diode (LD) as the device which converts the electrical signal into light. Both these devices are solid-state semiconductors which emit light at a predetermined wavelength.

Fig. 8.17, from Pirelli [1], shows the important differences in performance

A number of different wavelengths are simultaneously transmitted by one optical waveguide and on reception are separated by interference filters or diffraction gratings. Each wavelength can carry independent video, data or audio signals.

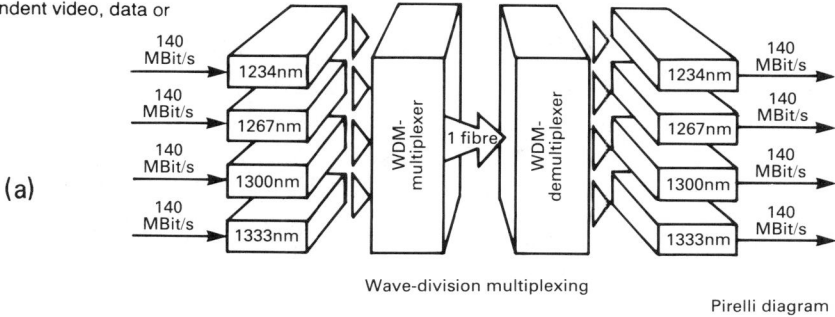

(a)

Wave-division multiplexing

Pirelli diagram

FEATURES

*   Good isolation.

*   Wide-band capability.

*   CATV, LAW, CCTV applications.

*   3-channel operation.

*   Operation within first window only (780 nm → 920 nm).

*   Mixed signal formats possible.

*   Tolerant to source wavelength variations.

*   Lightweight.

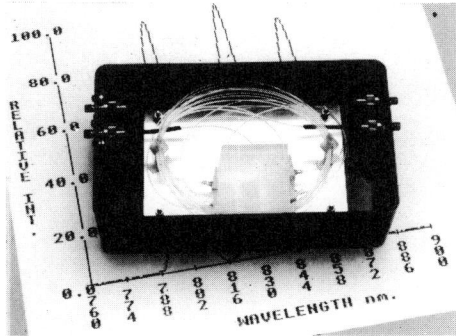

This Wavelength Division Multiplexer/Demultiplexer is designed to increase system flexibility and transmission capacity of an optical fibre system.

Selective narrow band optical filtering is used to achieve channel separation. The component may be used as a multiplexer and/or demultiplexer in conjunction with semiconductor laser diode sources emitting at 800 nm, 830 nm and 860 nm.

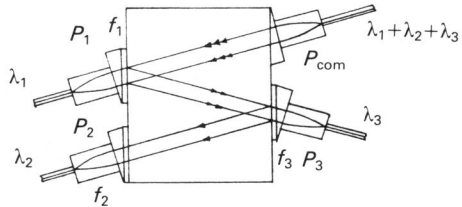

Optical isolation (dB):

Between outputs (common input.)				Between ports (as a multiplexer)		
λ(nm)	800	830	860	λ(nm)	800	830	860
800	–	31	43	800	–	60	60
830	30	–	40	830	60	–	60
860	36	30	–	860	60	60	–

(Typical measured results)

(b)

Fig. 8.16 — (a) WDM principle; (b) Marconi Wavlength Division Multiplexer (Reproduced by permission of GEC Research Ltd).

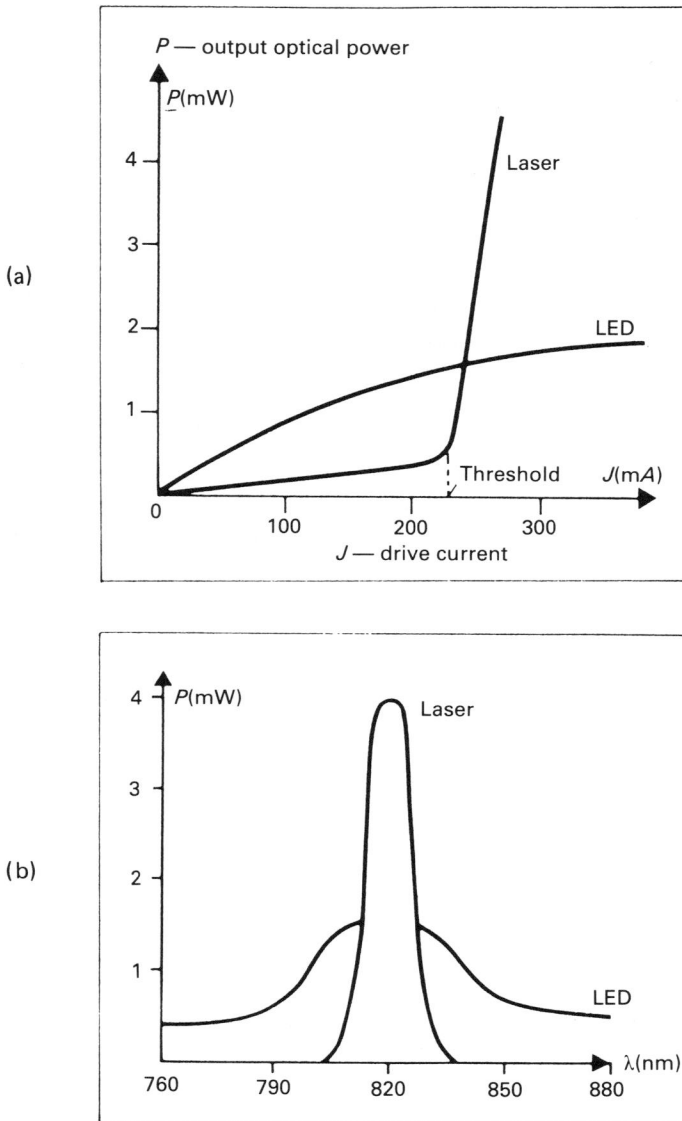

Fig. 8.17 — Performance comparisons of LDs and LEDs: (a) power/drive current LD and LED; (b) power/wavelength performance LD and LED. (Pirelli diagrams).

between a laser diode and a light-emitting diode, and it will be seen that the laser diode seems to have all the advantages.

The laser diode emits light of a single frequency and phase, with a narrow spectral width, perhaps as little as 3 nm, so that dispersion can be kept to a minimum; an LED is more likely to have a spectrum width of around 30 nm. The LD gives a higher power output than the LED for a given drive current, and is more linear over its operating range. Typical laser diodes might have output powers of 0.5–2.0 dBm (i.e. with reference to a milliwatt), whereas an LED might only produce around 18 dBm. The light output from a laser diode is more directional, i.e. given out over a narrower angle than that of the LED, which makes it easier to couple the light into the aperture of a fibre.

The only problems with laser diodes are that they are expensive to produce, require complex drive electronics, and usually need special cooling arrangements. Some lasers are not suitable for continuous duty, and are described as 'pulsed' lasers.

8.5.2   Laser safety
Although the optical power output of a laser diode may seem low, at typically about half a milliwatt, the concentrated high brightness of some laser beams can make them dangerous if viewed directly by the human eye, and precautions must therefore be taken to ensure that the safety measurements detailed in the codes of practice such as those published in IEC 825-84, BSI 4803 and ANSI Z136 1-80 are always adhered to. This will become even more important in the future, when it is expected that better lasers and improved coupling techniques between lasers and fibres will make it possible to make use of much higher launch powers. Increasing the launch power would have many advantages to the operator, providing far more flexibility in the layout of the system. Cable lengths between regenerators could be much longer, or the extra power could be used to allow cheaper, less sensitive optical receivers to be used. Alternatively it might be practicable to split the optical power between various distribution links, making for a more cost-effective system. At the present time it is possible to obtain, under laboratory conditions, a power launched into a fibre of about 110 mW from a laser of about 200 mW, and it is considered that such levels will be available commercially within a few years. The maximum power-handling capacity of a fibre-optic link depends upon many factors, including the length of the fibre, its non-linear performance capabilities, and the type of modulation to be used, but it seems likely that powers in excess of 500 mW will be feasible on long lengths of monomode fibre at wavelengths of about 1550 nm, and some experimenters think that several watts of power could be used on shorter lengths of fibre.

When considering laser power levels on fibres, it should be remembered that when wavelength-mutiplexing techniques are used it is likely that the combined optical output power could be relatively high, even though the power of each individual signal might be low, and appropriate precautions should be taken.

LEDs are very much cheaper than LDs, and are used mainly in short-range applications. They can be supplied already connected to a piece of fibre-optic, so that the 'pigtail' can readily be joined to the main length of fibre.

8.5.3   PIN diodes
The device used in the optical receiver to change the light back into an electrical signal must be very sensitive to changes in the incoming light signals, and must be capable of very fast operational response times if high data-rates, corresponding to wide bandwidth signals, are to be coped with. Virtually all the receivers currently in use with fibre-optic transmissions use one of two types of PIN (Positive Intrinsically Negative) diode (Fig. 8.18). These consist of a semiconductor p–n junction which is

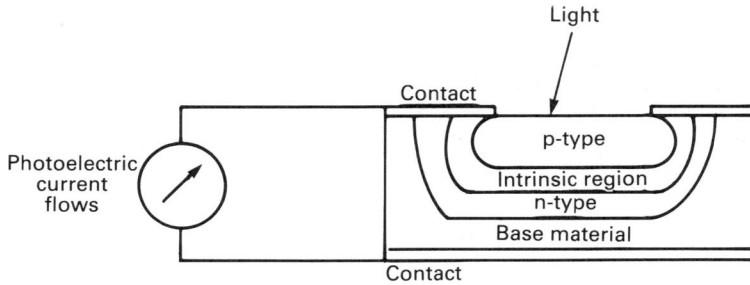

Fig. 8.18 — Basic construction of a PIN diode.

constructed in such a way that it can absorb the light energy falling upon it, and can then produce a photoelectric current which varies linearly with the intensity of the received light. PIN diodes are relatively inexpensive.

A development of the basic PIN diode is the avalanche photodiode (APD), which has the additional property of creating gain or multiplication of the electrons within itself, so that the APD is very much more sensitive than the simple PIN diode, and therefore suitable for the reception of signals over long-distance circuits. APDs are more expensive than PIN diodes.

Practical receivers can take many forms, ranging from simple die-cast boxes with optical input connectors and BNC video output connectors, to sophisticated multi-channel PCB modules. Fig. 8.19 shows a Pirelli multi-channel video receive/transmit unit designed for use on the monomode fibre section of a cabled distribution system, and it may be of interest to readers to see the brief specifications of this professional unit. It is designed to receive optical signals, convert them to electrical signals, which are then amplified and reconverted to optical signals for onward transmission. While the signals are in electrical form they may be extracted from the unit and fed to subscribers, so providing a tapping-off point from the main fibre-optic network.

8.5.4 Integrated optical devices
Although most current systems tend to carry out signal-processing operations in the electrical domain, much work is being done in research laboratories to make integrated optical devices which would do away with the need for many electro-optic conversion stages. We have already considered the optical-wavelength multiplexers that are now available, but although so far devices of this sort are only suitable for use on the trunk part of the network, and not for switching centres or subscriber drops, in the near future it seems that optical directional couplers and complex optical switches will become a normal part of fibre-optic systems. The Swedish Ericsson company has demonstrated an 8 by 8 optical switching matrix which can switch wide bandwidth signals quickly and with little loss, but does not expect to see commercial devices in use for another few years. Major developments in the provision of other practical, cost-effective optical processing devices can also be expected.

Experiments are being carried out by British Telecom with light-amplifier repeaters which will be able to directly amplify the light signals passing along a fibre-optic system, without having to first convert them to electrical signals. This form of

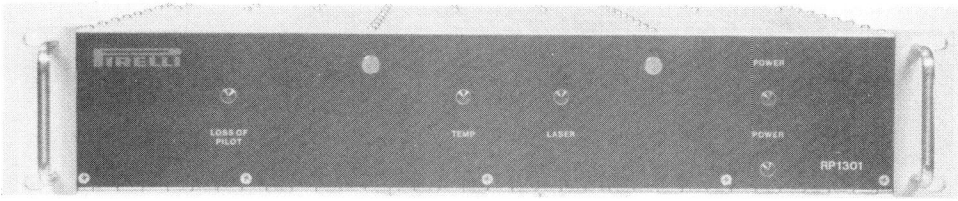

FEATURES

- Single Mode
- 1300 nm Wavelength
- 540 MHz Bandwidth
- Video Drop & Insert

GENERAL DESCRIPTION

The Pirelli RP1301 transceiver adds video drop and insert capability at any point on a Series 1300 FM fibre optic trunk or network. When combined with Pirelli 900 Series frequency agile FM equipment, these transceivers make new fibre optic system configurations possible.

The RP1301 transceiver is designed for receiving optical signals, converting them to electrical signals, amplifying them, and reconverting them to optical signals for retransmission. While in electrical form, signals can be dropped for local use, and additional electrical signals can be inserted for transmission.

The mainframe consists of five plug-in boards, a receiver, transmitter, interface and two power supply boards. The receiver contains a PIN receiver, an RF automatic gain control (AGC) amplifier module, and alarm circuitry. The interface board contains an RF amplifier that increases receiver output to the level required by the transmitter. The transmitter board contains a 1300 nm laser optical source, automatic laser current circuitry, and temperature control circuitry. This board also generates and retransmits the 10.7 MHz pilot tone used throughout the system for AGC and alarm purposes.

Alarm circuitry is provided locally on the power supply, receiver, and transmitter boards in the form of LED displays. The common alarms are dry-relay closures on the cabinet back panel.

Fig. 8.19a — Photograph of Pirelli optical transceiver, with specification. The specification is Fig. 8.19b.

light-amplifier uses a special solid-state laser crystal which can be persuaded to react to an incoming light signal on one of its faces by stimulating the internal production of a great deal more light, which comes out of the opposite face of the crystal.

At the University of Southampton, a team under Professor Alec Gambling is working on a project which would make the actual glass fibres 'lase' as the signal passes along them, thus stimulating amplification of the light signals as they travel along, doing away altogether with the need for separate repeaters.

Toshiba Corporation has developed an integrated optical demultiplexer which can separate up to five wavelength-multiplexed optical signals travelling along a fibre-optic cable into the original five individual signals (Fig. 8.20). Unlike conventional demultiplexers made of discrete components, the new unit integrates most of the key components onto a single crystal silicon substrate, which makes for a highly reliable and stable optical system.

The heart of the new receiver/demultiplexer is a light waveguide formed as a thin glass layer on the crystal silicon substrate. The five multiplexed light waves transmitted along the fibre-optic cable strike a refraction grating at a specific angle. The gratings then work like a prism, cleanly separating the multiplex into the five signals originally transmitted. The grating consists of a series of very fine rectangular grooves, 0.7 μm wide and 0.3 μm deep, etched on the substrate at 1.5 μm invervals.

These new types of optical device are at various stages of development, from extremely experimental to almost practicable, and the fact that such concepts are

TABLE A
RP1301 TRANSCEIVER SPECIFICATIONS

PARAMETER	SPECIFICATION
Receiver Optical Detector	PIN Photodiode, responsivity of .55 A/W @ 1300 nm
Receiver Optical Input	15 µW to 300 µW for constant RF output level
Receiver Distortion	None $<$50 dB below carrier level
Receiver Equivalent Input Noise Current Density	7 pA/Hz$^{1/2}$ from 10 MHz to 400 MHz

Transmitter Laser:
  Laser Life Expectancy
  Laser Type
  Optical Wavelength
  Laser Temperature
  Optical Bandwidth
	10^5 hours
	Injection Laser Diode
	1300 nm
	Stabilized at 20° C ($\pm$ 3° C)
	$\pm$ 3 nm, FWHM

Transmitter Optical Power Output	Not less than 0.5 mW average ($-$3 dBm)
Transmitter Distortion	All RF distortion products 40dB below carrier level at 50% modulation
Transmitter Noise	RMS signal-to-noise ratio of 50 dB or greater over the signal passband as measured in 4.2 MHz bandwidth
Alarms	N.O. contacts on back panel
	Receiver Loss of Pilot Alarm LED lights when pilot tone is not detected
	Transmitter Temperature Alarm LED lights when laser temperature varies $>$3° C from 20° C norm
	Transmitter Current Alarm LED lights when laser current varies $>$20% from factory setting
Pass Band	$\pm$ 1 dB, 20 MHz to 500 MHz
RF Input	Not $>-$1 dBm for single channel (derate for multichannels)
RF Output	For single channel, $-$10 dBm to 0 dBm (adjustable on the front of the receiver board)
Input/Output Impedance	50 ohms (75 ohms optional)
Fibre Pigtails	Single mode, 125 µm O.D. cladding, 9 µm core diameter
Optical Connectors	NEC D4 or equivalent (unless otherwise specified)
RF Connectors	BNC (unless otherwise specified)
Power Requirements	240 VAC @ 50 Hz 50 W
Size	19 $\times$ 17 $\times$ 3.5 inches (48.3 $\times$ 43.2 $\times$ 8.9 cm)
Weight	23 pounds (10.4 kg)
Mounting	19 inch mounting rack (48.3 cm)
Ambient Temperature	0° C to 40° C (32° F to 104° F)

Descriptions and specifications subject to change without notice.

Fig. 8.19b

currently being looked at indicates that completely new types of optical device are likely to have very exciting future potential, which should further stimulate the growth of the optic fibres as a communications medium.

8.6 PRACTICAL CATV SYSTEMS

Although theoretical fibre-optic cable bandwidths are much higher than those available on coaxial systems, and AT&T's Bell Laboratories have recently claimed to be able to carry 10 parallel 2 Gbit/s data streams, such fibres are not currently

Optical receiver

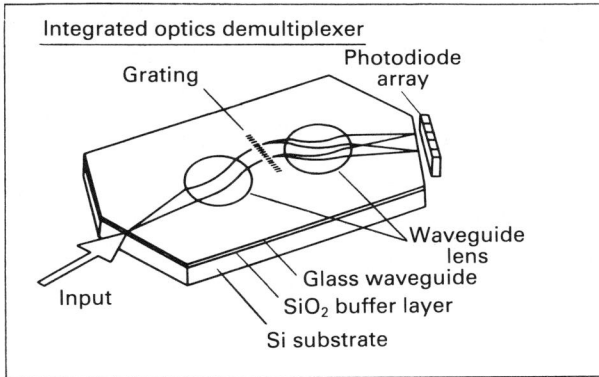

Optical transmitter

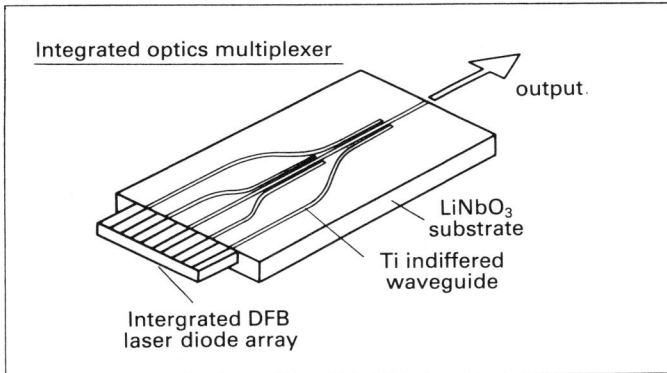

Fig. 8.20 — Toshiba optical multiplexer/demuliplexer.

available to the average cable operator who would like to use fibre-optics. Some of the very large cable systems that are being built in Europe involve very large capital expenditures, and are being built in cooperation with fibre-optic-equipment manufacturers, so that some of the most modern techniques discussed above can be incorporated.

Such advantages are not, however, available to the smaller operator, and he will have to make use of the much more limited equipment that is readily available. Fortunately the fibre-optic manufacturers have adopted a 'building-block' approach, so that all the cable operator has to do is to plan his system in a similar way to that which he would use for a conventional system, and then select his equipment requirements from the manufacturer's catalogue. In the same way that signal and power budgets are calculated before coaxial CATV systems are designed, optical-path-loss budgeting must be carried out before an optical system is built. Since the development of fibre-optic techniques is very rapid at the present time, however, it would be wise to make use of the fibre manufacturers' consultancy services, usually provided without charge, to determine the most appropriate building blocks for any system, in order to take advantage of the latest developments in this field.

Optical-loss budgeting requires knowledge of the performance of the various optical components, and if the calculations are carried out in dBm units, that is optical power relative to 1 mW, all the necessary calculations can be carried out using addition and substraction. Fig. 8.21 is an example provided by STC [2] of how such a

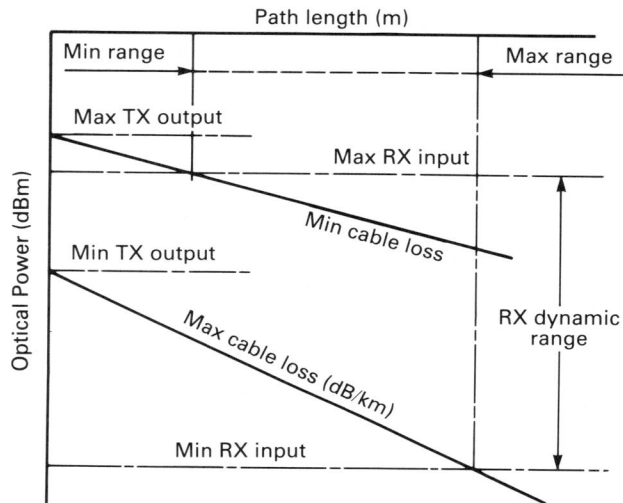

Fig. 8.21 — Optical-path-loss budget (Courtesy STC).

path-loss budget may be presented in graphical form, and is basically a graph plotting optical power against path length.

Fundamental design information required for path-loss budgeting
Cable: the maximum and minimum attenuation in dB/km at the wavelength to be used.

Connectors/couplers: maximum and minimum insertion loss.

Transmitter: maximum and minimum output levels at the wavelength to be used.

Receiver: minimum input level required for the specified signal-to-noise ratio required. Maximum input level before overloading takes place.

With the above information it is possible to mark the limits for maximum and minimum transmitter power and receiver-input sensitivity on the chart. The maximum allowable loss in the fibre can then be obtained by subtracting the transmitter power from the receiver sensitivity. Losses which will take place in connectors are usually taken account of by assuming that the transmitter power is reduced by the appropriate amount. The cable loss, which increases with length, is then drawn on the graph as the sloping lines for the maximum and minimum conditions, and the permissible operating conditions can be seen. It must be stressed that this is a much simplified budget in order to explain the principles, and that in real life, account has to be taken of the fact that the numerical apertures of the cable, receiver and transmitter are unlikely to be equal, and therefore additional losses will be sustained.

8.6.1 STC 'Multiview' system

The STC 'Multiview' system is typical of the type of equipment that is now readily available 'off the shelf' to the would-be CATV operator, and it will be seen that it permits the transmission of only four high-quality TV channels on each graded-index multimode fibre, and that these signals can only be sent a maximum of 4 km without repeaters (Fig. 8.22). This means that any operator wishing to carry a large number of channels needs to make use of several of the modules in parallel, and to send the signals along a number of separate optical fibres, which are in fact carried together in one small cable. Frequency-division multiplex is used to provide the four-channel capability, and the optical transmit and receive units can be used as repeaters if longer-distance transmission is required. All connections are made by using fibres that come complete with connectors, and using this type of approach, comprehensive CATV networks can readily be assembled.

Although it may initially come as something of a disappointment to learn that hundreds of TV channels cannot be sent over hundreds of kilometres using this type of system, there are still some very real advantages.

Optical fibres are immune to all forms of electromagnetic interference, which can and does from time to time cause havoc with coaxial-based systems. Once a coaxial system is carrying many different signals it is easy for the various signals to interfere with each other and to generate spurious signals and intermodulation products; this form of interaction cannot normally occur on optical systems. Whereas it is sometimes found that radiation from a conventional wired system can cause problems to viewers and listeners, there is absolutely no electromagnetic radiation that can cause interference from an optical fibre. It will be difficult for unauthorised users to tap into optical systems, whereas it is not unknown for some do-it-yourself electrical work to be performed on coaxial systems, a problem which should become less important as more secure scrambling systems come into use. For a given information-carrying capacity, optical fibres are very much smaller than coaxial cables, which should lead to better use of ducts and less expensive installations. One readily available cable of only 2 cm diameter contains 160 fibres in 'bundles' of 10 plus the necessary strengthening members, and another has 30 fibres and a steel tensile member within a 1 cm diameter.

One significant disadvantage of optical fibres is that they provide no direct-current path, so it is not possible to send electrical power along them to power repeater amplifiers, a technique that most cable operators currently use. For this reason some designs of optical-fibre cable also carry copper wires, but since repeaters are very widely spaced on optical systems it may not, however, be too difficult to provide local power sources at repeater stations, and most CATV networks are therefore unlikely to need this type of cable, which will be much more expensive than standard fibre-optic designs. In many practical applications it will, however, prove necessary to have cables that are strong enough to withstand being pulled through

General description

STC Components' FDM fibre-optics video system permits the transmission of up to four high-quality TV channels on each graded-index multimode fibre, providing a minimum unified weighted signal-to-noise ratio of 50.5 dB on each channel with an optical path loss of up to 28 dB (in 50/125 μm multimode fibre).

Design of the FM transmission modules is based on narrow-stripe 850 μm CW semiconductor lasers manufactured by STC Components Laser Unit, Paignton, and commercial avalanche photodiode detectors. 850 nm wavelength operation has been selected owing to the ready availability and cost effectiveness of optoelectronic components and fibre at this wavelength. However, the modular nature of the equipment will readiliy permit the introduction of longer-wavelength devices, single or multimode, as and when requirements dictate.

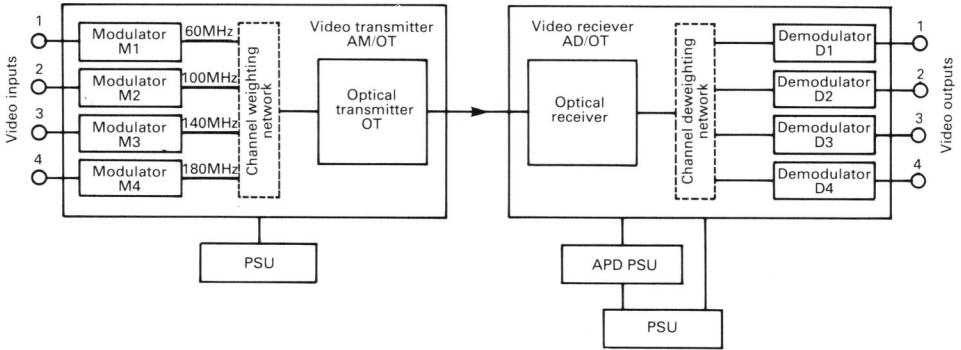

The modules are constructed as plug-in double-height Eurocards measuring 233.4 mm (6U)×220 mm, width 25 mm (5E) and are based on a number of 'building blocks' to enable a wide range of configuration to be provided. Each component building block is referred to by a one- or two-letter mnemonic describing its function: viz: FM Modulator — M; Optical Receiver — OR; Optical Transmitter — OT; FM Demodulator — D.

Examples of existing combinations include: 4M/OT — 4-channel modulator/optical transmitter; 4D/OR — 4-channel demodulator/optical receiver; OR/OT — wideband optical repeater. In addition to other specialised combination, two non-optical functions are provided: APD PSU — avalanche photodiode HT invertor; FOA — fan-out amplifier (1 in — 8 out). More detailed specifications on all the available modules are included in separate data sheets.

System configurations

STC MULTIVIEW provides high-quality repeaterless COTV performance over distances up to 4 km using 850 nm graded-index multimode fibres featuring losses of 3 dB/km and dispersion bandwidths of 600 MHz-km. Greater transmission distances can be achieved for lower-quality surveillance video, or by the use of intermediate repeaters.

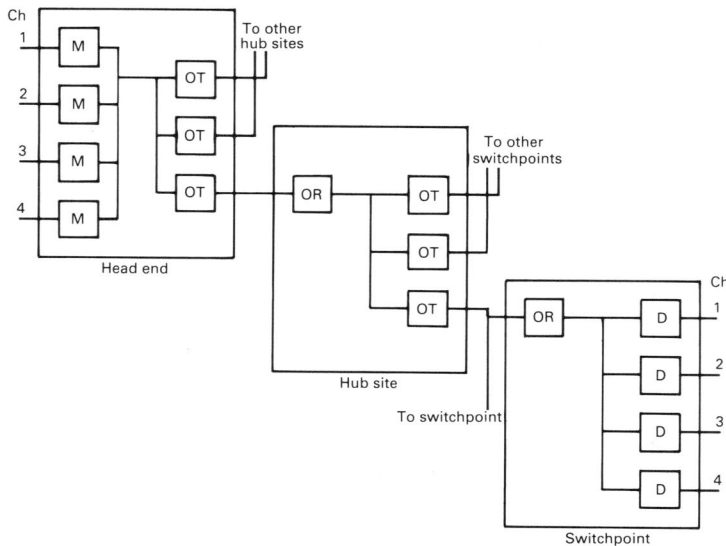

Modules 4M/OT and 4D/OR are directly compatible with any CCITT video interface (IV pk-pk, 75 Ω) and can be configured into a 4-channel video link using a minimum of additional equipment (APD PSU module and mains PSU), and without any specialised skills or expertise. Optical connection to both optical transmitter and receiver modules is easilyaccomplished using fibres terminated with BT-approved Stratos single-way connectors.

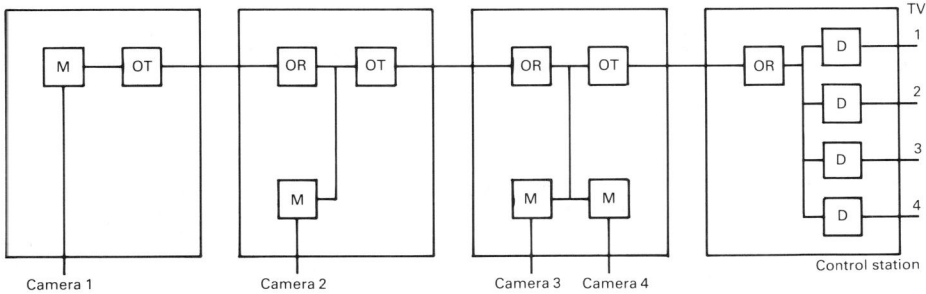

Extended range operation can be achieved by the simple expedient of including broadband (i.e. without demodulation to baseband video) repeater module (OR/OT) at appropriate points in the optical-fibre link. Further video channels may also be injected at these intermediate points, making the network ideal for distributed surveillance systems (e.g. motorway surveillance).

Fig. 8.22 — A practical fibre optic CATV system, The **STC Multiview** Modular System.

ducts, so strength members have to be added, which can be of metal or a synthetic material such as Dupont's Kevlar, which combines strength with light weight. Many fibre-optic trunk cables will be armoured for extra protection. In total contrast to this vision of very heavy cables, British Telecom has been experimenting with persuading their very fine monomode fibres to travel along narrow ducts by blowing them in using compressed air, a method that has been found to work very well in certain situations.

There is no doubt that the overall performance capabilities and the scope for future developments make fibre-optic systems the choice of the far-sighted engineer, but as in all things we cannot ignore the price factor. So many of the costs are as yet unknowns, and it will be fascinating to see how the balance between coaxial and fibre-optic systems develops where it most counts, in the marketplace.

REFERENCES

[1] Pirelli Publication OS/02/04/86. Optronic Systems.
[2] J. D. Archer, *STC Manual of Fibre Optics Communication.* 5191/2571E Ed. 4.

9

Two-way cable systems — interactive services

So far we have considered only signals that have been travelling from the head end to the subscribers' premises, and such signals are called 'downstream' signals, or, alternatively, are said to be following the 'forward path'. One of the major advantages offered by cabled systems over conventional terrestrial transmissions or even satellite transmissions, which could turn out to be the key factor in ensuring the successful growth of enhanced cabled systems in the future, is that a permanent physical link exists between the head end and the subscriber. This means that it should be possible, providing that the network topology allows, for the subscriber to make and maintain two-way contact with the head end, for the purpose of sending messages, instructions, or various other kinds of signal to the head end. The transmission path followed by signals travelling away from a subscriber's premises towards the head end is known as the 'reverse path', and signals travelling in this direction are known as 'upstream' signals.

Any cable system which incorporates both forward and reverse paths is known as a 'two-way' or 'bi-directional' system (Fig. 9.1).

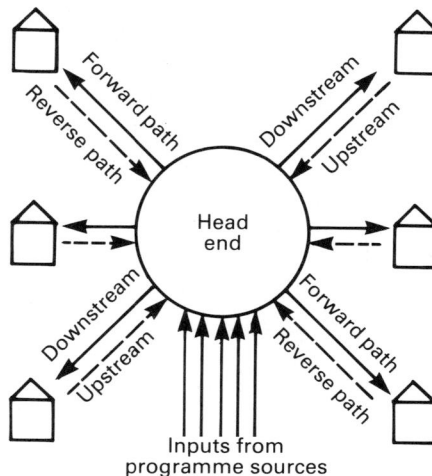

Fig. 9.1 — Forward and reverse paths.

The word 'interactive' is defined in BS 6513:8.02 as meaning 'capable of mutual action, e.g. between a subscriber and a service provided on a cabled distribution system', and a note then says that this sort of capability will usually be provided only on a two-way system.

Having taken a rather pedantic look at the terminology, we shall now go on to consider just what is involved in a two-way cable system, what changes are needed to the standard equipment, and what advantages two-way and interactive systems can provide.

It is not strictly necessary for an interactive service to have the capability of transporting wideband information, and a simple form of interactive cable service could be envisaged where the 'reverse path' was formed by a standard telephone circuit, so that the operator at the head end could be requested, for example, to play out a particular videotape of the customer's choice through a particular cable channel. Although this is not what is generally meant by an interactive cable service, the example does serve to remind us that a great deal of the two-way information that customers need can be conveyed by narrowband circuits such as the telephone line; services such as videotext can provide access to vast quantities of data stored in databases throughout the world, without the requirement of anything other than a standard telephone channel with a fairly restricted audio bandwidth.

At the time of writing, the standard UK dial-up public switched-telephone network (PSTN) can cope fairly comfortably with data-rates of up to about 4800 bit/s, and with the more expensive type of error-correcting modems that are starting to appear on the market at reasonable prices it is even possible to send and receive data signals at 9600 bit/s. Once the modern 'System X' exchanges are installed more extensively over the next few years, the introduction of an Integrated Services Digital Network (ISDN) will allow the existing standard telephone lines into customers' homes to carry data at up to 144 kbit/s, a data-rate that should allow for virtually all of a domestic or small-business customer's information needs, as well as being able to provide high-quality still-television pictures if complex bit-reduction coding algorithms now being worked on are used. This could, for example, allow a customer to call up any one of thousands of still-pictures via the telephone line, perhaps to allow the visual selection of houses from an estate agent's photographic file, or to browse through the electronic equivalent of today's mail-order catalogue.

With possibilities like these available from standard telephone lines, and some 86% of UK homes having a telephone, it might be thought difficult to establish the need for wideband interactive cable services for the domestic user, who only really needs wideband capabilities if he wishes to send moving television pictures from his home to some other place. The enormous amounts of digital data which need to be sent to represent a moving television picture, and the rates at which these data must be sent, ranging from 216 Mbit/s to perhaps 34 Mbit/s, make it unlikely that existing telephone lines will ever be used for this purpose, whereas any suitably designed modern cable system can easily cope.

Whether or not present-day cable system customers can perceive the need to send moving pictures upstream, the capability will exist, and history tends to show that people eventually find all sorts of uses for new developments provided by techno-logy. The tendency for more people to work from home, whilst staying in touch with all their normal office services via some form of telecommunications link, commonly known as 'teleworking', may well lead to the need for the average customer to be able to send lage amounts of data quickly, perhaps sending high-speed facsimiles of prepared documents back to the office, or detailed technical drawings to a customer. Business users have already woken up to the importance of being able to transmit data quickly around the world, and interactive cable systems will allow the business-man to make use of similar 'professional' data-transfer systems from his own home.

The UK Westminster Cable TV company already provides a 'films-on-request' delivery system which makes use of banks of laser video disc players situated at a central point, but although this is one of the most technologically up-to-date systems it should be noted that the upstream messages, i.e. the requests for particular video discs to be played, need not make use of any wideband capabilities of the system.

9.1 UPSTREAM STANDARDS

It might at first sight be thought that the technical standards for upstream signals could be more relaxed than those for the main downstream signals being distributed over the whole network. This is not in fact the case, because the standards committees decided that they must take account of the possibility that a two-way cable system could be used in such a way that television signals from an 'outside broadcast' point could be introduced into a convenient part of the network. These pictures would then travel upstream to the head end, from where they would be distributed downstream throughout the whole cable network. This means that some of the standards for upstream transmission must be even tighter than those for downstream signals, and examples of this can be seen in British Standard 6513. A case in point is the carrier-to-random-noise ratio for CCIR system I television signals (i.e. as used in the UK) where 43 dB is the permitted minimum downstream [2], whereas 50 dB is required for the upstream case [3].

9.2 PRACTICAL TWO-WAY TRANSMISSION SYSTEMS

We have already seen that the actual network topology is probably the single most important factor in determining whether interactive services can be made available. Tree and branch networks, although providing a good cheap distribution network, are generally unsuitable for use as communications networks, and the switched-star layout makes for a far more practical 'two-way' cable system, as well as being well suited to fibre-optic construction techniques.

At the present time, no recognised optimum method has evolved for the provision of fully interactive services using fibre-optic cables to and from the subscriber's home, although options such as frequency-division multiplexing, wave-length-division multiplexing, and even multiple optic fibres in one cable are being seriously considered. It can be seen however, that fibre-optic systems will be able to use similar two-way techniques to those that are already well established on coaxial-cable-based systems, which will now be considered in some detail.

9.2.1 Two-way system techniques

We have seen that a standard coaxial system is capable of carrying a wide range of television and radio signals over a bandwidth of perhaps 400 MHz or more, using carefully planned frequency-division multiplexing techniques. The most commonly used technique for allowing upstream signals to be carried is to reserve part of the available bandwidth, usually at the lowest-frequency end of the available spectrum, for the upstream transmissions. Fig. 9.2 shows one way in which the spectrum is

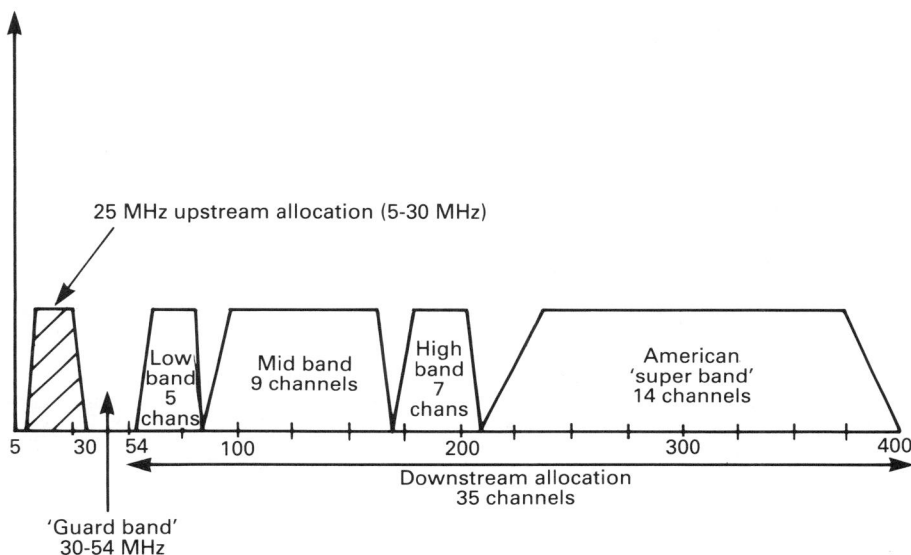

Fig. 9.2 — Typical spectrum allocation for upstream and downstream use. The midband channels shown are the European standard. The American midband channels start at 120 MHz and end at 174 MHz (as shown extending to the high band channels).

divided up for this purpose in some American networks, and it will be noted that care has been taken to avoid using any of the standard VHF channels used for downstream transmissions. Although it would be theoretically possible to put the upstream and downstream frequency bands closer to one another by the use of carefully designed filtering arrangements at the head end and in the receivers, this would prove too expensive to be predictable in many cases, so that it is usual to allow a wide 'guard band' between the upstream and the downstream services, as Fig. 9.2 shows, even though this could perhaps be considered wasteful of spectrum space.

The system shown allows for some 25 MHz of spectrum bandwidth to be used for upstream transmissions, but there are as yet no firm standards set for this, and it is perfectly possible to use other spectrum layouts if it becomes important to allow for the carriage of greater numbers of upstream signals.

As another practical example, British Cable Services 'System 8' provides for a single upstream video channel on 39.5 MHz on one cable, and for 250 kbit/s of data transmission in the band from 2 to 4 MHz on several cables, as well as allowing for smaller quantities of data to be carried in the band below 10 kHz. in addition to this, part of the band from 2 to 4 MHz has been reserved for system control data, and packet data can be sent in the band up to 10 MHz, for electronic-mail-type services. The band limits for this could be extended to 25 MHz if future demand indicates the need.

9.2.2 Design considerations

It should be noted that calculations for noise, distortion and echo performance have to be carried out in both directions, which can make the design of two-way networks a fairly complex affair. It is normal to design the downstream layout first, so that the number of amplifiers in the chain will be known. As we saw earlier, the downstream system will be designed so that amplifiers are spaced in accordance with the losses of the cable, couplers and taps. This means that the upstream layout may not be anything like ideal, and since the cable losses at the low frequencies used for the upstream signals will be very different from those used for the downstream signals, there are often tremendous differences in the gain requirements of upstream and downstream amplifiers at particular points along the system, and it can be very difficult to maintain the correct operating levels at all the upstream amplifiers. In practical systems the gain of the upstream amplifiers tends to have to be high to overcome the losses introduced by splitters, etc., in the downstream path, and this in turn leads to the need for attenuators to be inserted at the upstream-amplifier inputs in order to keep operating levels correct.

In a downstream system the noise contribution from the head end will be small, but it is important to note that noise in the upstream system is not only that due to the number of amplifiers in the cascaded system back to the head end, but also has a component that depends upon the total number of upstream amplifiers. All upstream amplifiers are effectively contributing noise towards the system head end, since they all direct their signals towards it, and it is sometimes helpful to think of each upstream amplifier as a source of noise, or even as a noise generator. It then becomes clear that the amount of noise from each generator would add to that from every other generator at the head end.

In any practical system it is most unlikely that every subscriber, or even many subscribers, would be making use of the upstream capabilities of the system at the same time. Some systems even use what is known as an 'interrogation' technique, which only allows one customer at a time to send signals to the head end, under the control of the head-end computer. If those subscribers who are not currently using the upstream facilities could have their upstream equipment temporarily disconnected from the system, this would reduce the number of 'noise generators' contributing noise towards the head end, and would therefore improve system performance considerably. Such techniques are now possible, and the use of 'return-feeder disconnect' may well prove to be essential in large interactive systems, although the complexity is unlikely to render the technique viable in small systems.

Upstream cross-modulation can, however, be calculated exactly as for the downstream case, since it will depend on the total number of amplifiers through which the signals pass. The usual care must be taken to check that signal levels and amplifier inputs and outputs are kept within the design limits.

Simple one-way amplifiers must give way to more complex units containing input and output filters and two separate amplifiers, one for each direction of transmission, throughout a two-way network, and Fig. 9.3 shows the essential components of such

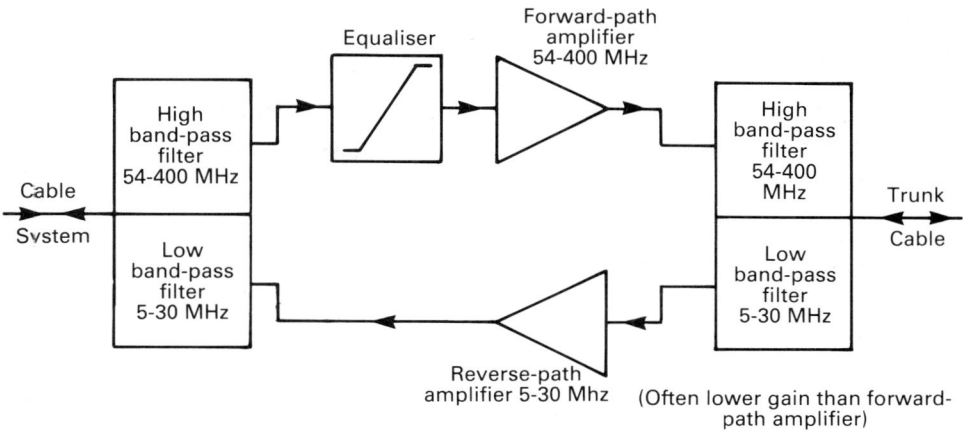

Fig. 9.3 — Schematic of a simple two-way cable amplifier.

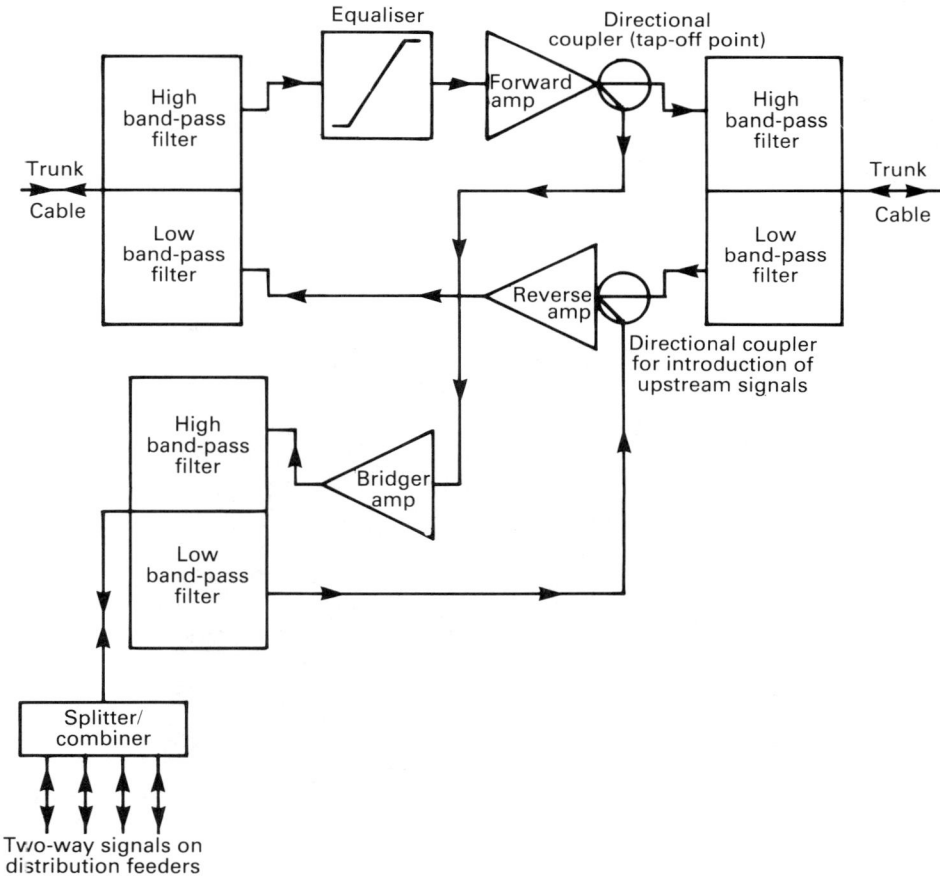

Fig. 9.4 — Schematic of a two-way mainline/bridger amplifier combination.

a unit. Practical amplifier designs now on the market have provision made for the plug-in connection of a second amplifier module for the upstream signals, and for the necessary filters.

We saw earlier how mainstream/bridger amplifier combinations are frequently used in one-way systems to amplify the main trunk signals, whilst tapping off some of the signal and amplifying it before sending it at a higher level along distribution feeders. Exactly the same technique can be used for two-way systems, but it will be seen from Fig. 9.4 that the additional bi-directional filters and amplifiers make the unit rather complex, especially when the necessary automatic gain control circuitry and power insertion filter circuits, which are not shown on the diagram for the sake of clarity, are added.

9.3   UPSTREAM SIGNALS

Although we have seen how a system can, if necessary, carry television pictures and sound upstream, most of the uses for two-way systems so far envisaged involve the transmission of varying amounts of digital data. As well as the customer being able to send requests to the head end for different programmes to be delivered, he can ask the head end to connect him to remote computer databases to obtain information, perhaps for telebanking or teleshopping purposes, and can then use a personal identification number and credit card to transfer funds between his bank and the seller of the goods (Fig. 9.5)

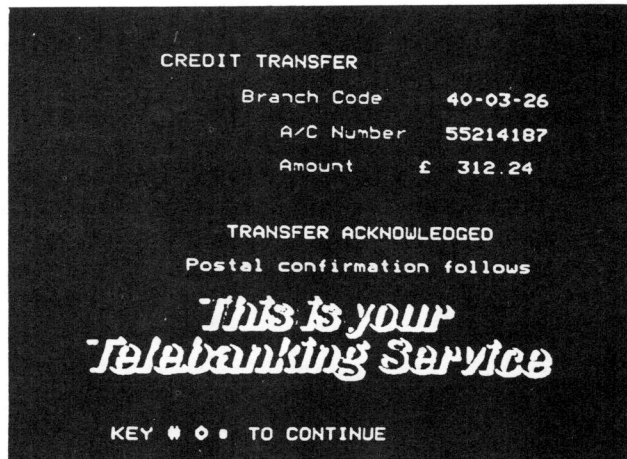

Fig. 9.5 — Telebanking screen (Photo courtesy BCS).

Almost incidental from the customer's point of view, but vital from that of the operator, is that interactive systems allow for accurate and flexible billing of the customer, since charges can be recorded on the head-end computer at any time that a premium-chargeable service is requested. The fact that there is a physical connection between subscriber and the head-end controller means that the source of any request can be identified, and that the subscriber can therefore be charged for the services with which he is supplied.

The upstream channels can also be used to provide information from the subscriber to the cable operator as to which channels are being viewed, and other audience-measurement information. This type of information can enable the operator to obtain accurate statistical information about the number of viewers who are choosing to watch particular channels, and can assist him to make financial decisions based on these statistics.

A small step from this is to have electronic opinion polls where subscribers to the cable network can send back their answers to questions posed in television programmes shown on the system, giving their opinions on anything from the credibility of a politician's speech to the pop singer most likely to succeed on the local talent show. Small-scale trials of this sort of thing have been carried out in many different parts of the world, from the CUBE system in Columbus, Ohio, to the Hi-Ovis system in Japan, and in the UK, some audience reactions to television programmes are collected in a similar way. If we ever reach a stage where virtually the whole population has access to this means of feedback, we could hold frequent national electronic referendums on a wide range of public-interest topics, enhancing the private citizen's contribution to democracy, and keeping the government on its toes.

Data sent back to the head end on a two-way network can include information on the operational status of the actual network, by sending monitoring signals back to the head end from each amplifier, allowing the location of any problems to be quickly determined. It is now possible to send digital data corresponding to many measured parameters of the television signal back to the head end, which can provide constant monitoring of the system, and it seems reasonable to foresee future systems where this information is made use of to control a completely automatic network, where spare amplifiers would automatically be switched into service by a computer if it detects something wrong at a particular point in the network.

Many other services are being introduced on two-way cable systems, and it may well be that these are just as important as extra programmes to the successful implementation of cable systems, since the customer can be shown the advantages of being connected to a cable system that competing programme delivery services such as satellites and terrestrial transmitters just cannot provide.

Remote betting on horse racing, dog racing, etc., is already proving very popular in many systems, and when these services are linked with direct viewing of the sporting event concerned, many more viewers will decide to subscribe to the services concerned. The system used by British Cable Services allows for related videotext messages to be superimposed on the sporting pictures, and this type of development could allow the customer to place his bets remotely, without taking his eyes off the race!

Fire- and burglar-alarm sensors in the customer's home can be connected to the cable system, to provide continuous monitoring of their status at the head end. The head-end operator can identify where the alarm originates, and immediately contact the police or fire brigade. Even in systems that cannot provide a full wide-band upstream capability, slow-scan television cameras could be used to keep a watchful eye on subscribers' premises, sending pictures to the head-end operator, and it is possible for any movement showing up in an otherwise stationary surveillance picture to trigger an alarm that attracts the operator's attention.

Remote meter reading is likely to prove a popular option with electricity, gas and water utilities once enough people are connected to cable systems to make such schemes economic. The advantages to the authorities concerned are obvious, but the customer also benefits in that he does not have to wait at home for a meter reader to call.

Electronic-mail services and messaging services can be provided by means of interconnections between the head end and the various national computer communication services, and even local telephone services can be provided free of charge between customers connected to the cable system. If the present British Telecom/Mercury duopoly position on the British telephone system can be relaxed, it may be possible for cable subscribers in different towns to obtain cheap telephone services by means of interconnections between different systems.

As a first step along this road, Cabletime Limited is experimenting with the integration of standard telephone services, from Mercury Communications, with its switched-star cable networks (Figs 9.6 and 9.7). The telephone signal cables are kept quite separate from those belonging to the cabled distribution network up to the point at which the 'drop' cable leaves the switch in the street to enter the viewers home. Combination of the telephony and television signals takes place at this point, giving the advantage that just one co-axial cable into the home provides the viewer with telephone services as well as the usual radio and television programmes. A standard wall socket has separate outlets for these services. It may seem surprising that the telephone cables are kept separate from the main feeder cables on the network, when it could have proved possible to multiplex the telephone signals onto the main cable, but the fact that this less sophisticated approach has been adopted gives yet another indication that what seems the 'obvious' technical solution to a problem may in fact give rise to financial or organisational problems, and it may well be that the telephone company felt that it was necessary to keep their cables separate from the main cable until the connection is made into the home.

Home computers have proved very popular in the UK, and some cable operators are offering the facility of downloading software for these computers (telesoftware), enabling the subscriber to 'pay per play' with computer games, or to pay for instructional or utility programmes which are sent along the cable on request.

It may well turn out that it is the wide range of 'extra' services that cable systems can provide, over and above the television pictures, that will persuade people to subscribe, and it is not unreasonable to envisage that a local 'gossip' network, a sort of electronic parish magazine, could become so successful that people would feel that they were missing something if they did not join the cable network.

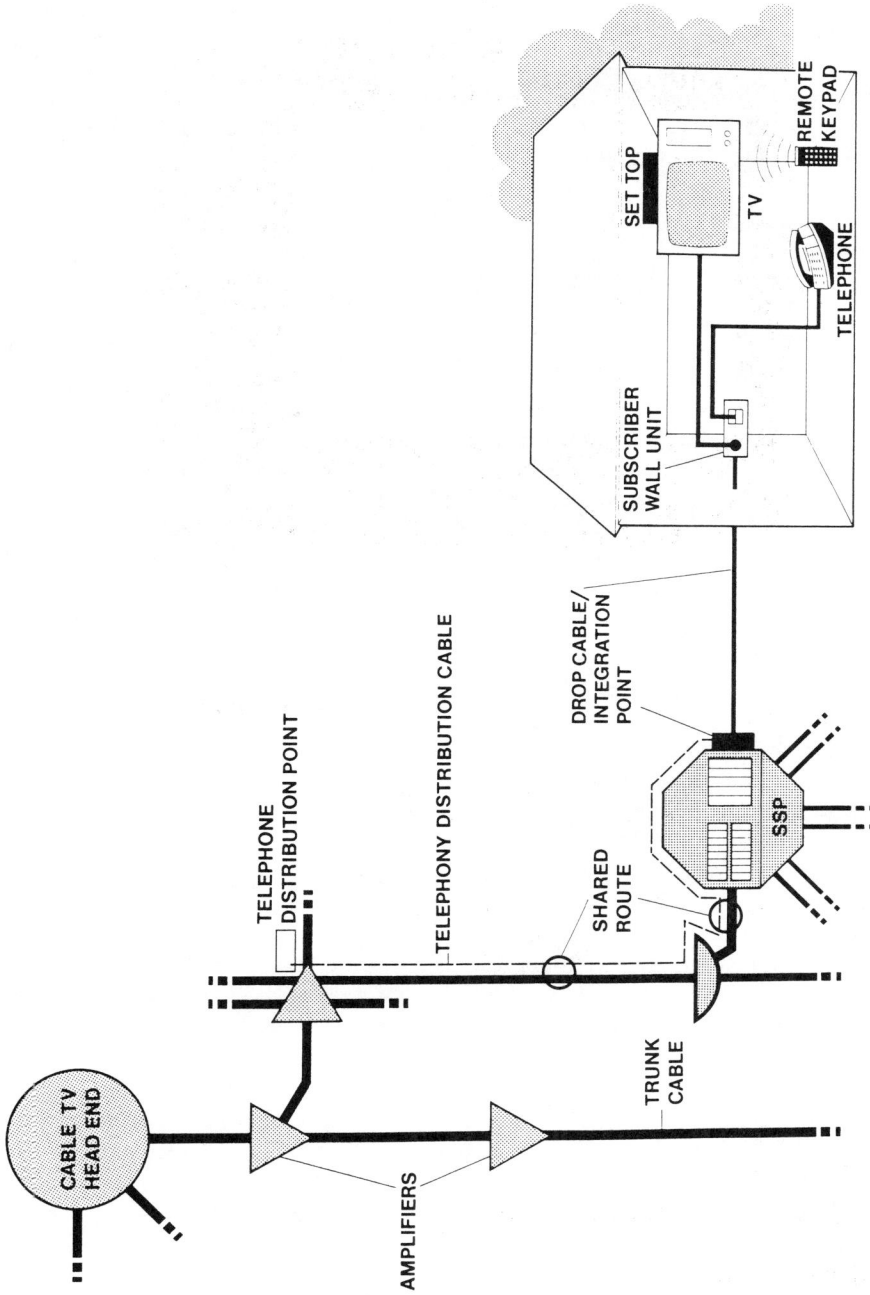

Fig. 9.6 — Drop cable integration of telephony on cabletime switched star system (SSP).

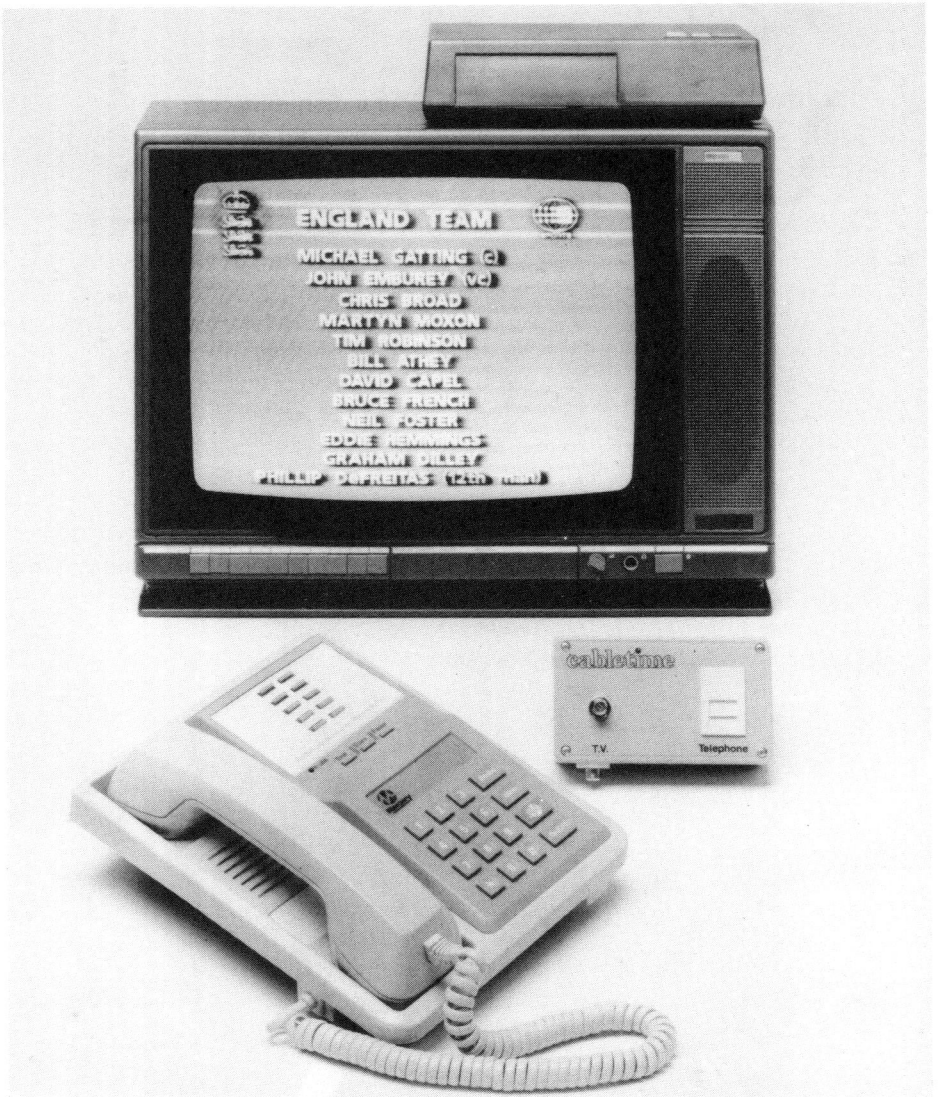

Fig. 9.7.

REFERENCES

[1] CCIR Report 624. Characteristics of television systems.
[2] BS 6513:3.2.6.
[3] BS 6513:4:2.4.

10

Satellites and cable — a powerful combination

The advent of powerful Direct Broadcast Satellites (DBS) in the late 1980s will inevitably lead to satellites being thought of as direct competitors to cable services, but this has not really been the story so far, and television distribution satellites are providing, and will continue to provide, an invaluable source of extra programmes for the cable operator and his customers (Fig. 10.1). The competition with satellites

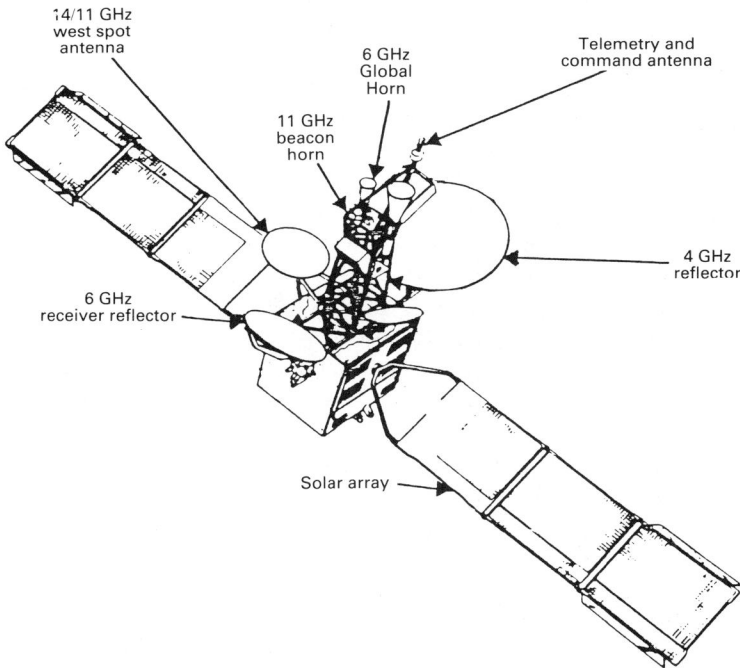

14/11 GHz west spot antenna

6 GHz Global Horn

11 GHz beacon horn

Telemetry and command antenna

4 GHz reflector

6 GHz receiver reflector

Solar array

Fig. 10.1 — Intelsat V — a typical TV distribution satellite (photo from Intelsat).

will come about because the strong satellite signals will mean that viewers wanting a wider choice of programmes will for the first time be able to decide to install their own small dish aerials, perhaps as little as 35 cm in diameter, and pull in signals from broadcasting stations all over the continent. The worry for cable operators is that customers will prefer to buy this kind of satellite-receiving equipment, rather than subscribing to a cable service, especially if the costs of the dish aerial and receiver are kept low. Customers will then have to be convinced that subscribing to a cable service still gives them practical advantages, and it will therefore be even more important than it is now for the cable operator to keep his customers happy by supplying them with a wide range of attractive programmes at a reasonable price. In addition, however, the cable subscriber must be offered extra services, and we have seen already how the various interactive services are able to provide a whole range of new facilities, which could be the deciding factor as to whether the viewer chooses to subscribe to a cable network.

The cable companies in the United States and in Canada pioneered the use of satellites for programme distribution in the early 1970s, generally using satellites that had originally been intended for telephone traffic, although the experimental Hermes satellite, launched in 1976, showed that it would be technically feasible to have higher-powered satellites that could be used either for direct home reception or for community use. The Canadian 'Anik' satellites are used to bring television signals to remote communities in the far north of the country, where they are frequently received on large dish aerials and then sent via cabled distribution systems to the viewers' homes, although it is not unknown for some viewers to receive the signals directly on their own 'backyard' dishes. This 'eavesdropping' on services intended for cable operators is something that has become more widespread throughout North America, Europe, and other parts of the world as the continuing increase in power transmitted by the satellites enables viewers to pick up the signals on smaller dishes. Unauthorised reception of this kind is leading to problems of copyright that looks as though they will result in most satellite programmes eventually having to be scrambled to prevent abuse.

To give some idea of the tremendous resource that distribution satellites can provide for the cable operator, there are currently some 18 different satellites radiating over 40 different television programmes that a cable operator in North America can make use of. In Europe, an operator who is prepared to equip his head end with various dishes of diameters of as litttle as 3–5 m can receive about 20 different programme services from five satellites, and plans are well advanced for this number to be significantly increased in the near future. Most of these are currently transmitted in the clear, but some services are scrambled, and the cable system owner has to make arrangements with each service operator if he is to make use of the programmes, which can lead to difficulties when many different services are to be received. Not only are there problems in actually contacting the appropriate source of the programme, but different scrambling methods are used by different satellite distributors, which can mean that several separate decoders need to be leased or purchased. To try to overcome these difficulties in the UK, an organisation has been set up called the UK Cable Programme Provider Group, which has decided that it will recommend all its members to standardise on one scrambling system. After studying various possibilities it has agreed that this can best be done if its members will use the MAC (Multiplexed Analogue Components) system for all transmissions, because of the sophisticated scrambling and customer-addressing facilities that are built into the specification of the MAC system.

Although there have been many problems reported with rights' agreements between actors, agents and satellite operators, these do not concern the cable operator, who must, however, ensure that he has the permission of the satellite-service provider before distributing the various programmes.

10.1 SATELLITE HEAD-END EQUIPMENT — TVRO — SMATV

These days, even the smallest of cable systems can boast its own satellite-receiving equipment, and this type of receiving station is often known as a TVRO, or Television Receive-Only station. The term SMATV (Satellite Master Antenna Television) is used to describe small wired systems serving blocks of flats or apartments which have a satellite-receiving antenna as well as the normal MATV-receiving aerials. The satellites which are used for television distribution purposes are invariably in a circular orbit some 36 000 km above the equator, at which altitude they circle the earth in a period equalling that of the earth's rotation. This means that as far as an observer on earth is concerned, the satellite will appear to remain in the same place in the sky, and for this reason the orbit is known as the 'geostationary orbit' (Figs 10.2 and 10.3). The main advantage of this orbit for our purposes is that

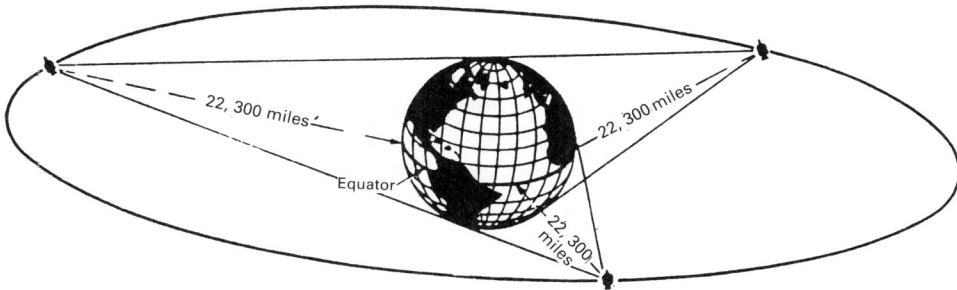

Fig. 10.2 — Geostationary orbit (Courtesy IBA).

we can use a fixed receiving aerial directed at the satellite, and we will not require a complex moving antenna to track the satellite across the sky. The satellite will, however, occasionally need to be slightly reorientated, which is done by small on-board rocket motors. This only takes place for a few minutes every few weeks, and under normal circumstances the cable operator will not be aware that any change is taking place.

10.1.1

Distribution satellites of this type are equipped with several 'transponders', receiver/transmitter combinations which receive signals from a ground station on earth via

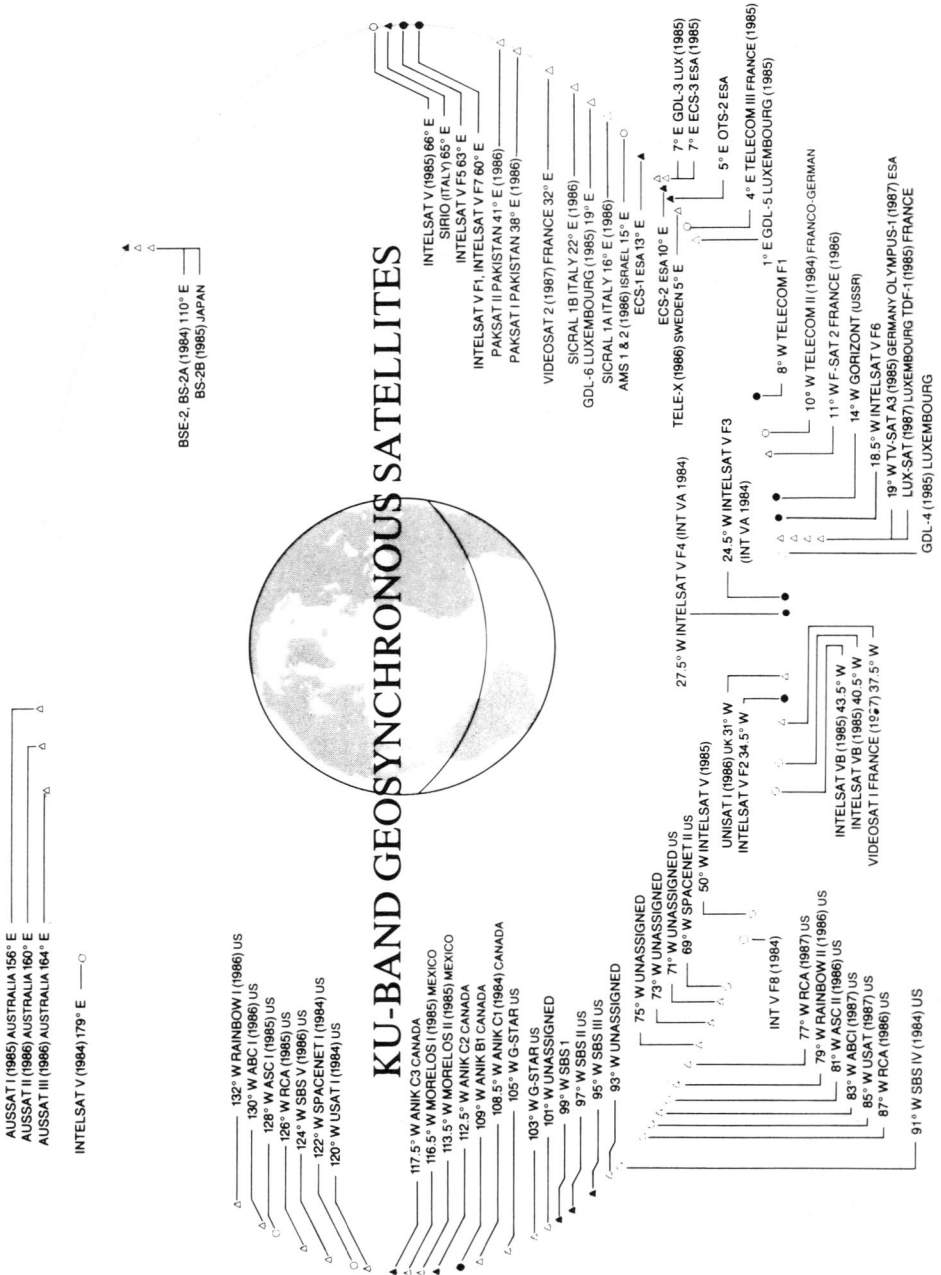

Fig. 10.3 — Television satellites around the geostationary orbit (Reproduced by permission of *Satellite World*, March 1985).

what is known as the 'uplink', then carry out a frequency change to prevent interference with the uplink, and then transmit the same programme signal towards earth. This signal transmitted through the 'downlink' is usually arranged to cover a particular part of the globe, western Europe or the east coast of the United States, for example. This is convenient, since coverage areas, sometimes known as 'footprints', can be tailored to suit particular needs by using directional aerials, and the gain achieved by these aerials allows the satellite's limited power to be concentrated on the desired reception area, giving a higher power flux density, which allows modest receiving aerials to collect enough power to provide satisfactory pictures. Typically, a distribution satellite will have a transmitter output power of about 20 W (i.e. approximately 13 dB above 1 W, usually expressed as 13 dBW). When this signal is concentrated into a spot-beam by the antenna on the satellite which might typically have a gain of 35 dB, we end up with an Effective Isotropic Radiated Power (EIRP) of around 13 dBW + 35 dBW = 48 dBW, which is some 70 kW.

This term EIRP, the product of the power supplied to the satellite's transmitting aerial and its gain, is commonly used in satellite literature, and information about the power available from a satellite at any given point on the earth's surface is frequently given in the form of EIRP footprints like the one shown in Fig. 10.4.

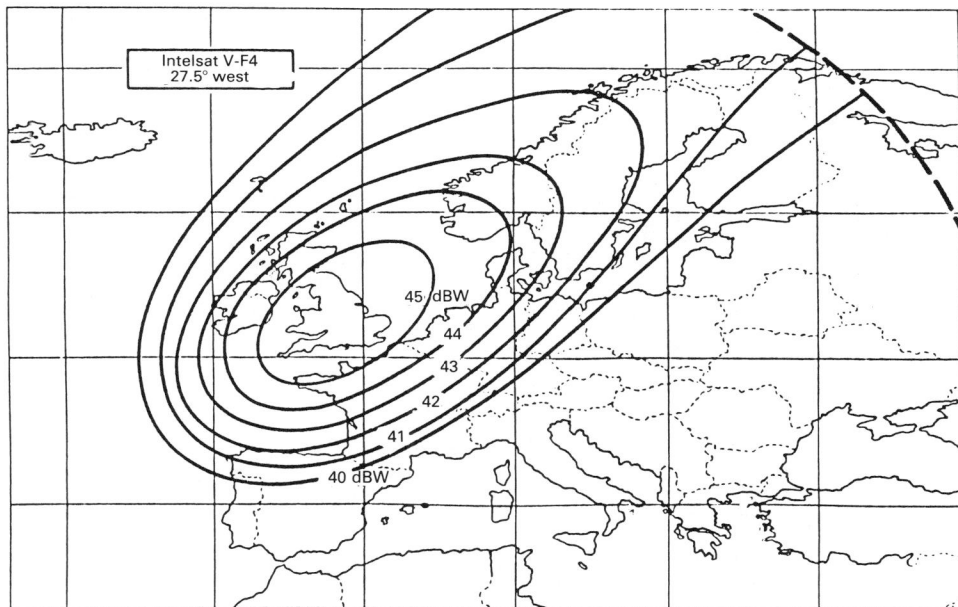

23

Fig. 10.4 — EIRP footprint for Intelsat V. (Intelsat).

With a knowledge of the EIRP, a power budget for the link between satellite and receiver can be calculated. Knowing the noise performance of the receiver, the bandwidth of the signal and the efficiency of the antenna, the appropriate size of dish antenna for the wanted carrier-to-noise ratio can then be obtained. Further details of the technology of television satellites of this type, and of the calculation of detailed power budgets can be found in ref. [1], but the example below shows the sort of figures that are involved when receiving a service typical of those designed for cable operators in Europe, Super Channel. This is radiated from transponder number 12 of the European Communications Satellite ECS-F1, and for the purposes of our example it is assumed that the signals are received on a 3 m dish by a cable operator in Brussels.

Satellite: ECS-F1 at 13° east.
Frequency: 11.674 GHz, linear (vertical) polarisation.
Sound: + 6.65 MHz from vision.
FM deviation: 25 MHz/V.
Video bandwidth: 5.5 MHz.
RF bandwidth 36 MHz.

TVRO location: Brussels, Lat. 51.00   Long. 4.50 east.
Receive aerial elevation angle 31°, azimuth 10° east of south.
TVRO antenna shock: 3.0 m.
Low noise block (LNB) amplifier/convertor noise figure 2.4 dB (effective noise temperature 211 K).

Distance to satellite 38 519 km.

Satellite EIRP	46 dBW	(i)
Free space (spreading) loss	205 dB	(ii)
Mean attentuation due to astmosphere	0.5 dB	(iii)
TVRO aerial gain	49 dB	(iv)
TVRO aerial efficiency	65%	
TVRO aerial 3 dB beamwidth	0.55°	
TVRO aerial coupling losses	0.78 dB	

Total signal power available at TVRO site = (i) − (ii) − (iii) + (iv)
$$= -100.6 \text{ dBW}$$

LNB noise temperature	211 K
Antenna noise temperature	65 K
Total system noise temperature	276 K
System figure of merit (G/T)	25 dB/K
Receiver noise bandwidth	36 MHz
Carrier-to-noise ratio	18 dB
FM improvement factor	36.77 dB
Video signal-to-noise ratio (weighted)	54.77 dB

The appropriate satellite EIRP can be obtained from the footprints published by satellite operators, and the required elevation and azimuth angles for the receiving aerial can be obtained from charts drawn up by the equipment manufacturers. Thus although it is important for a cable operator to check in advance that his proposed head-end satellite-receiving equipment will give satisfactory results, the various charts and tables that the equipment manufacturers provide can be used to keep mathematical calculations to a minimum.

10.1.2   C band and Ku band

In America, most of the television distribution satellites operate in what is usually known as 'C band'. IEEE standard 521–1976 [2] defines the C band as 4000–8000 MHz, but invariably the uplink frequency is at about 6 GHz (i.e. 6000 MHz) and the corresponding downlink is at about 4 GHz. As examples, a cable operator wanting to receive signals from Intelsat IV, located at longitude 1° west, would pick up signals from the BrightStar network on 4.065 GHz, and programmes intended for the American Forces on 4.175 GHz. In Europe, most television distribution satellites work in the so-called Ku band (nominally 12.5–18 GHz), which means that uplinks are at about 14 GHz and downlinks at about 11.5 GHz. The cable operator who wishes to provide his customers with the greatest possible choice of programmes will therefore need to equip his head end with several dishes pointed at the satellites of interest, and this may well involve the installation of both C-band and Ku-band equipment, and will also necessitate aerial feeds which can work on different polarisations, to suit the transmitted signals, since various different polarisations are used on different transponders to obtain the best possible use of the limited number of frequencies available, with the minimum of interference between satellites.

On some satellites, including the 'European' Intelsat at 27.5° west, so-called 'half-transponder' operation is used to increase the number of programmes that it is possible to broadcast simultaneously. Two programmes can share a single transponder without intermodulation problems if the power and deviation of each of the two programme signals are reduced. Normally each television signal is reduced in power by about 3 dB, with the result that the effective isotropic radiated power from the satellite is about 4 dB below that permissible with just one programme carrier.

Table 10.1 gives just a few examples showing that different satellite transmissions have different characteristics, and the cable operator who wishes to provide the widest possible choice of programmes for his subscribers must ensure that the satellite-receiving equipment which is purchased can cope with all these types of signal, and with the different types of scrambling that the programme providers may use.

Just to receive the small selection shown in Table 10.1, the cable operator will need at least five dishes pointing in different directions, with the ability to select different signal polarisations. Three different types of descrambling equipment would also be needed, as well as a special receiver for the MAC signals, that will include a decoder for the digital sound which this system uses. A further complication is that the low-noise block amplifier/convertor units usually sold cover the frequency range from about 10.9 to 11.7 GHz, which means that it will also be necessary to buy a separate, different LNB to receive the signals from the French Telecom B satellite which are above 12.5 GHz.

Cable system operators contemplating the purchase of TVRO equipment should also notice that if they are eventually intending to relay the true DBS signals to their

Fig. 10.5 — Photograph of typical TVRO head-end receiving equipment.

customers, these programmes will be radiated in the band between 11.7 and 12.5 GHz, so that yet another type of LNB will be required, as well as special receivers for the multiplexed analogue component signals that will be used.

10.1.3 TVROs — cable operators can take the lead in satellite reception
Although the purchase of all this TVRO equipment might seem daunting to all but the largest of cable operators, the many different satellites in use can work to the cable operator's advantage. The average DBS user is not going to want to buy more than one set of satellite-receiving equipment, which will restrict his choice of new programmes, whereas the well-equipped cable system will be able to offer a very wide range of satellite-borne programmes at little extra cost. This should put the cable operator in a very competitive position, especially if he can use advertising to spell out the advantages of being a subscriber to the cable system before the viewer succumbs to the blandishments of the DBS company. This is especially important to cable operators in the UK who know that they have a short period of grace before the UK DBS services begin; in the intervening period, subscribers can be shown that

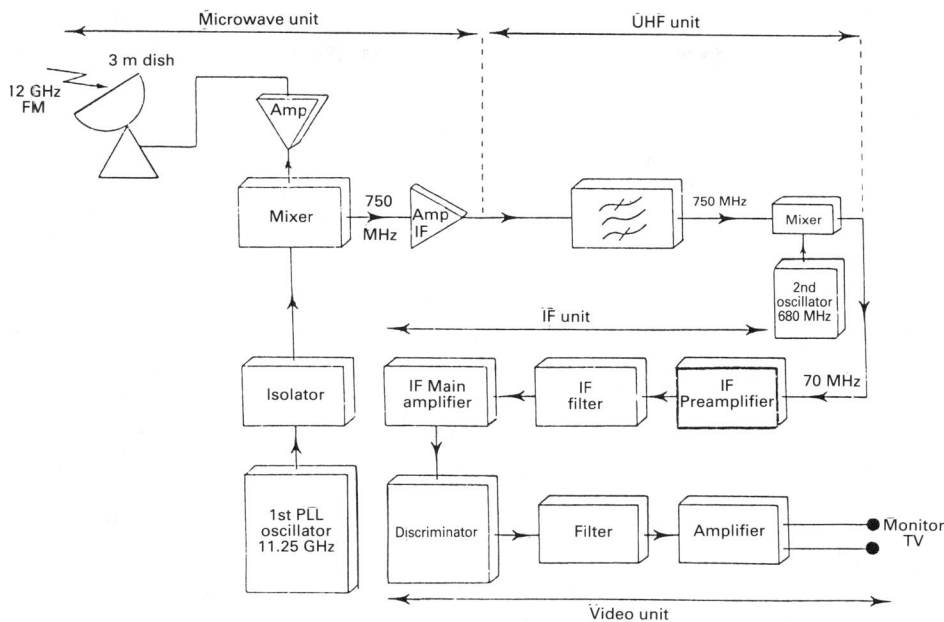

Fig. 10.6 — Block diagram of equipment required at head-end for television reception from medium-power distribution satellites.

Table 10.1

Programme and frequncy	Satellite and location	Sound carrier spacing	Transmission standard	Polarisation	Scrambling
Arts Channel 11.135 GHz	Intelsat V 27.5° E	6.6 MHz	PAL	H	Clear
Sky Channel 11.550 GHz	ECS-F1 13° E	6.65 MHz	PAL	H	Oak Orion
NRK 11.180 GHz	ECS-F4 10° E	Digital sound	C–MAC	H	Clear
La Cinq 12.506 GHz	Telecom 1B 5° W	5.80 MHz	SECAM	V	RTC Discr.
Filmnet 11.140 GHz	ECS–F1 13° E	6.60 MHz	PAL	V	Matsushita
Eins Plus	Intelsat V 60° E	6.65 MHz	PAL	H	Clear

satellite programmes are available from the cable 'now', whilst being reassured that the eventual DBS services will be provided as they become available, without the need for the customer to buy or rent any new equipment.

10.2　MAC (MULTIPLEXED ANALOGUE COMPONENTS) — A BROADCASTING STANDARD OPTIMISED FOR BOTH CABLE AND SATELLITE

Although the NTSC, PAL and SECAM colour television systems have worked well for many years, it is important to remember that the major factor that the developers of these systems had to bear in mind was that the introduction of colour had to have as little impact as possible on the millions of black-and-white receivers that were in use at the time when colour was to be introduced. The methods that were used to overcome potential non-compatability problems have been well documented [3,4], and there is no doubt that the various colour systems were successful in meeting their objectives, but inevitably some compromises had to be made in these so-called 'composite' colour systems, where colour was added to black-and-white pictures by the addition of a suitably modulated subcarrier.

Fig. 10.7 shows how the pioneers of the composite colour systems discovered that

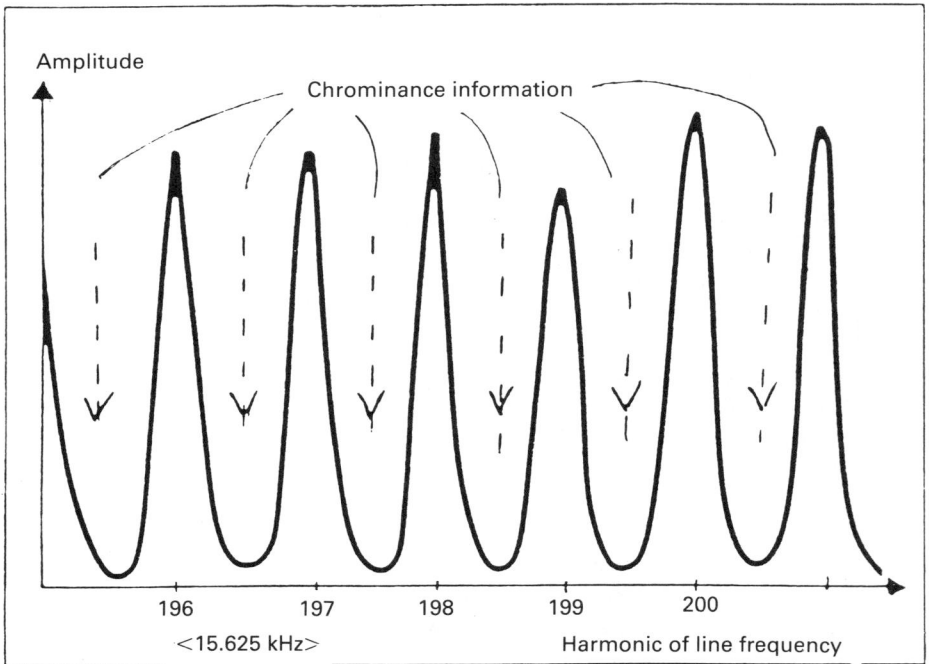

Fig. 10.7 — Energy spectrum of 625-line television signal.

the energy spectrum of a 625-line television picture consisted of a series of peaks occurring at harmonics of the television line frequency, with gaps between. It was the realisation that these gaps could be used to carry the colour information that made compatible colour and monochrome broadcasts possible, and the result was a frequency-division multiplex system where the colour and monochrome (black-and-white) frequencies overlapped to some extent.

Fig. 10.8 shows that the resulting baseband spectrum of a PAL signal has the

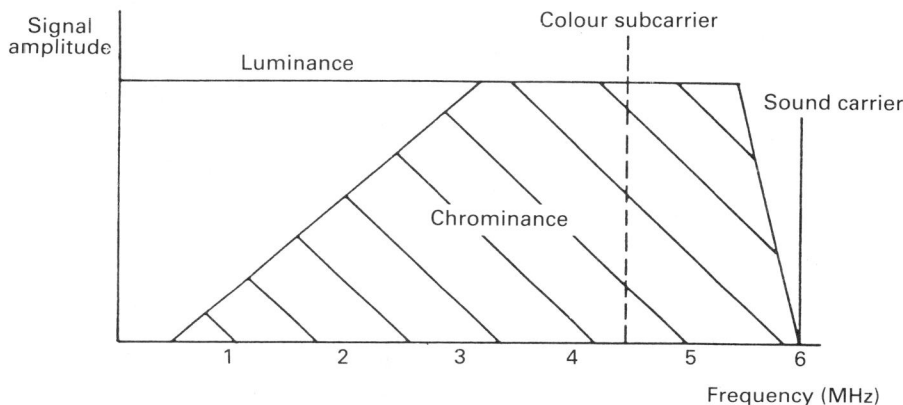

Fig. 10.8 — Baseband spectrum of a PAL signal.

colour information mixed with the luminance information, which can make it very difficult for a receiver to separate colour from detailed (i.e. high-frequency) black-and-white signals, and this gives rise to two undesirable problems known as cross-colour and cross-luminance. Cross-colour is the most noticeable effect, and is familiar to most viewers as the appearance of spurious coloured patterns when a television announcer wears a checked suit or fine-striped shirt. These effects were known to the pioneers of the composite colour systems, who considered them negligible, but they have become more noticeable as the high-frequency-performance capabilities of receivers and studio equipment have improved over the years. Broadcasting engineers in the 1980s started to realise the advantages that could be obtained by using colour signals in the form of their individual components, initially Red, Green, and Blue, but later the luminance signal Y and two colour difference signals, U and V, and the idea was born of a broadcasting system that would overcome the problems of the earlier composite systems. The introduction of direct broadcasting from satellites gave the opportunity to introduce such a component system for transmission, since everyone who wanted to receive the DBS broadcasts would need to buy some form of adaptor box to add on to their existing receivers, and it was realised that these adaptor boxes could contain the circuitry required to decode the new type of broadcast signals.

Satellite transmissions need to use frequency modulation to make the best possible use of the limited power that is available on spacecraft because of practical limitations on the physical size of the solar-cell arrays. It is well known that FM transmissions have a triangular noise spectrum, as shown in Fig. 10.9, which implies

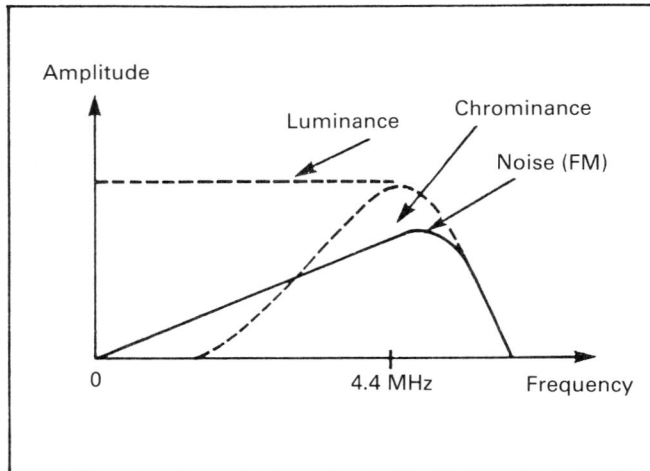

Fig. 10.9 — Triangular noise spectrum of FM signals.

that the system noise will be concentrated in the high-frequency parts of the spectrum. Since PAL pictures carry their colour information in the high-frequency parts of the picture spectrum. PAL signals carried over FM channels are subject to chroma noise, which is particularly objectionable in large areas of saturated colour.

To overcome these problems, research engineers at the UK Independent Broadcasting Authority developed a completely new system of transmitting television pictures, the Multiplexed Analogue Components (MAC) system [5]. The term 'multiplexed analogue components' strictly describes just the vision signal, which consists of the analogue chrominance components (the colour information), and the analogue luminance component (the black-and-white information) of the picture, which are transmitted during separate periods of time using a technique known as 'time-division multiplex'. Fig. 10.10 shows one line of the MAC signal, and it will be seen that since the luminance and chrominance signals are never transmitted simultaneously, there is no chance of cross-colour or cross-luminance arising.

To enable the separate chrominance and luminance signals to fit into a standard 64 μs television line period, which is essential if compatibility with existing television receivers is to be maintained, both signals are time-compressed on transmission in order to pack them into less than 64 μs, and once they reach the receiver they are expanded so that both the black-and-white and the colour parts of the picture once again fill a complete active television line. Care has been taken to control the amounts by which the colour and monochrome components are time-compressed, so that the system noise is fairly evenly spread over the whole picture spectrum, thus

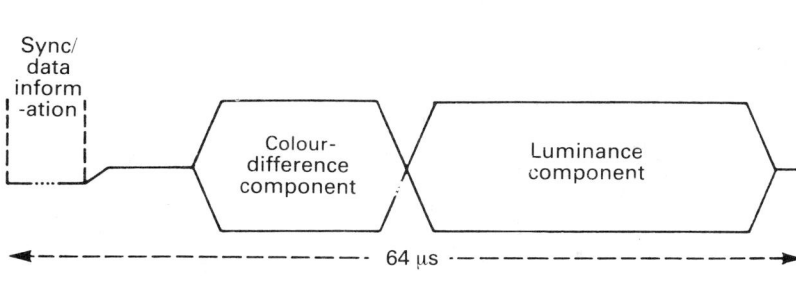

Fig. 10.10 — One line of a MAC signal showing how luminance and chrominance ae transmitted separately.

eliminating the problem of FM PAL signals having all the noise concentrated in the saturated-colour parts of the picture.

It will be seen from Fig. 10.11 that a period of about 9 μs in the line-blanking interval before the transmission of the colour-difference signals is given over to the transmission of synchronisation information and data. These data are carried at the very high rate of 20.25 Mbit/s in 'packets' of 751 bits, and can provide numerous high-quality sound channels as well as a considerable amount of auxuliary data. Since this burst of data at 20.25 Mbit/s occurs only for a short period at the beginning of each line, this means that on average just over 3 Mbit/s of data can be carried along with the picture, and some of this capacity can be used to carry encryption information for conditional-access subscription systems. Various ways of adding this data to the basic MAC signal have been proposed, and the European Broadcasting Union has agreed that a 'family' of MAC systems having different data characteristics can be used for satellite broadcasting and for transmission over cable systems.

10.2.1   D-MAC

The system known as 'D-MAC' has been chosen by UK government for transmission from satellites, and it is also eminently suited to simple conversion for use on cable systems. 'D' stands for 'duobinary', a three-level digital-coding system which can carry the 20.25 Mbit/s data in a base bandwidth of about 8.5 MHz, and we saw in Chapter 7 the details of how such a multi-level coding system works. Satellite broadcasting systems have a transmission bandwidth of about 27 MHz, which means that a direct FM D-MAC signal would tak up too much valuable spectrum on the average cable system, so that for cable use the operator is likely to demodulate the received FM signal from the satellite and then remodulate it on to an AM vestigial sidebands carrier at VHF and UHF. Practical tests have shown that the resulting D-MAC AV/VSB signals can be carried on cable systems where the channel spacing is around 10.5 MHz, and ref. [6] gives full details of the excellent results that were obtained on two very different types of cable system at Swindon and at Kingston in the UK.

Although much time and energy was given to negotiations aimed at getting a

Fig. 10.11 — Off-screen photograph showing the time multiplex of the compressed chrominance and luminance components of a MAC signal, together with the data burst (Courtesy IBA).

single satellite standard for Europe, The French and German DBS services will use a modified form of D-MAC, known as D-2 MAC. This uses the standard MAC TDM (time-division multiplex) waveform for its picture information, and also uses a duobinary data format, but the data-rate is reduced to 10.125 MHz, half that used in D-MAC. This means that only half the number of potential sound channels of D-MAC can be used, but the advantage is that duobinary data at 10.125 MHz can, if necessary, be carried in a base bandwidth of only just over 5 MHz. When these D-2 MAC signals are remodulated onto an AM/VSB carrier they take up just over 7 MHz of bandwidth, which fits in well with many continental cable systems which use channel spacings of this order, although some of the high-frequency picture information is also lost (Fig. 10.12). To obtain the full-quality D-2 MAC picture signals in

System	FM (nominal)	VSB/AM	
		Min.	Full quality
C	27	15.0	15.0
D	—	10.5	10.5
D2	27	7	10.5

Fig. 10.12 — Comparison of D-MAC and D-2 MAC bandwidth requirements for FM satellite transmissions and for AM/VSB cable transmissions. (All figures in MHz)

AM/VSB format, a total cable bandwidth of about 10.5 MHz is still needed, and many engineers think that it seems shortsighted to adopt a less-than-optimum system, throwing away half the potential data capacity, for the sake of compatibility with old-fashioned cable systems, many of which will probably be replaced with wider-band systems within the next decade.

10.2.2   Standards for component signals on cable
IEC Standard 728 was prepared before the possibility of carrying multiplexed analogue component signals had arisen, but appropriate study groups are currently considering the modifications that will be required. It seems likely that such parameters as maximum and minimum carrier levels at system outlets, permitted carrier-level differences, amplitude and phase response and frequency stability will be assigned different tolerances according to the type of signal being distributed. Provisional parameters for cable systems wanting to carry the FM satellite signals directly, after frequency downconversion to perhaps 900–2000 MHz, are also likely to be included, to allow for possible future developments, although no currently planned systems are intending to use this technique.

10.2.3   Enhancements
One of the major advantages of the adoption of a MAC system is that the systems

have been designed in such a way that in the near future various enhancements are possible, allowing viewers the option of higher-definition pictures with wider, more cinema-like aspect ratios. These options will be provided in a completely compatible way, so that existing viewers continue to receive their normal pictures, whilst those who are prepared to buy more-sophisticated receivers will be able to have wider aspect ratio pictures with improved subjective vertical definition and the complete elimination of interline flicker. If the data capacity of the vertical blanking interval is included with that of the normal MAC data burst, about 4 Mbit/s of data can be sent along with the pictures, some of which could be used to carry information intended to tell special processing units in future receivers, known as adaptive upconvertors, how to generate an optimal higher-quality picture. Such 'digitally assisted television' techniques would enable broadcasters to transmit truly compatible standard- and higher-quality television services simultaneously, allowing viewers to choose between different types of receiver for different needs. The field of enhanced services is another area which existing IEC regulations for cable systems do not cover, but study groups are looking at the implications that these enhanced signals will have, and cable operators who have engineered their systems to carry full-bandwidth D- or D-2 MAC AM/VSB signals should encounter no particular difficulties once the enhanced services become available.

10.3   SOUND BROADCASTING FROM SATELLITES

As has been discussed, the MAC system makes provision for up to eight high-quality sound channels to accompany each vision signal, and it seems likely that since not all of these will be required to provide television sound, some of the available MAC data capacity will be used to carry radio programmes on the standard 12 GHz satellite signals. Cable operators who send the demodulated data signals into each customer's home will have no problems, since the MAC decoder in the home will cope with the data decoding for each of the required sound channels, but operators who choose to decode the sound signals at the head end and remodulate them may find that the large numbers of sound carriers required become prohibitive.

Cable operators wanting to install systems that are 'future-proof' should note that the West Germans are currently planning to introduce a 'sound-only' satellite service intended for reception at cable head ends. This service is likely to use QPSK (Quadrature Phase-Shift Keying) modulation which will be able to provide up to 16 stereo sound programmes from just one satellite transponder. German cable networks are planning to carry these sound programmes at frequencies of between 111 and 125 MHz, so special cable receivers will be required.

Broadcasters in other countries in Europe are also keen to see the establishment of radio broadcasting from satellites, but the highly directional aerials needed to pick up reasonable amounts of signal at 12 GHz make the normal satellite band unsuitable for reception on anything other than fixed installations. Much discussion is going on to try to secure the use of a band of frequencies in the range 0.5–2.0 GHz, which would make reception on portable receivers practicable, and the 1988 World Administrative Radio Conference will be asked to make some spectrum available for this purpose. Cable operators wanting to provide all possible services for their subscribers should therefore keep an eye open for the outcome of these talks, although it is likely to be quite a few years before such services begin, even if a suitable block of frequencies is allocated.

REFERENCES

[1] J. N. Slater and L. A. Trinogga, *Satellite Broadcasting Systems*, Ellis Horwood, Chichester.
[2] Radar Frequency Bands, IEE Standard 521–1976.
[3] H. V. Sims, *Principles of PAL Colour Television and Related Systems*, Iliffe Books, 1969.
[4] P. S. Carnt and G. B. Townsend, *Colour Television*, Iliffe Books, 1969.
[5] Specification of the MAC/packet family of systems, EBU Tech 3258.
[6] H. J. O'Neill and P. A. Avon, The distribution of MAC on cable, *Cable Television Engineering*, **13**, no. 7, December 1986.

11

Conditional access — scrambling and encryption

Although we have so far concentrated on the technical aspects of cable television, it is important to remember that the installation and continued operation of any cabled distribution system will depend on the availability of finance. Whereas many local councils have, as a service to their ratepayer, provided the capital cost of the necessary CATV equipment to receive the basic television services provided by the broadcasters and ensured that maintenance is carried out at little cost to the viewer, this type of operation has really depended upon the fact that the programmes being relayed were available free 'off-air' to the cable operator, since the costs of the broadcasters had already been covered by the receiving licence fee or advertising. Once extra programmes such as feature films become available, however, their providers need to be paid, and some means has therefore to be found of financing these programmes. The simplest answer is to use the cable network to carry advertising on these extra channels, but although there is no doubt that the amount of advertising can be increased to provide some extra revenue, it seems unlikely that this revenue will continue to grow to an extent sufficient to finance the dozens of different channels that are likely to be available to cable subscribers over the next few years.

Opinion-poll data and practical experience over a number of years have shown that some viewers are prepared to pay a premium to receive extra services, especially if the new service can offer something different from the main channels, perhaps recently premiered cinema films or live sporting events such as world championship boxing matches. If revenue is to be generated from such programmes, some means has to be found of rendering the programmes unwatchable to those who do not wish to pay, whilst enabling those subscribers who *are* prepared to pay to watch their chosen programmes. Although it would in theory be possible to achieve this discrimination between the two different kinds of viewers by simple switching on a perfect star-distribution system, where each viewer is directly connected to the head end, we have already seen that such systems would be completely uneconomic, and it is therefore usual to encode or scramble the signals radiated from the head end, providing those viewers who choose to pay with an appropriate descrambling box to restore the pictures to normal.

Such arrangements are known as subscription-television, pay-television, or toll-television, and there are two main types of system. The first type of subscription-television is where the viewer pays a regular subscription in order to be able to receive programmes on a particular channel. The second type is known as 'pay-per-view', and involves the viewer in choosing to pay for individual programmes as they are shown. This usually means that some signal has to be sent from the viewer's home to the head end to indicate that the viewer wishes to watch the current programme; the head-end operator can then send an enabling signal to the viewer's receiver to switch on the descrambling circuitry and allow the viewer to watch. Before sending the enabling signal, the operator will check to see that the subscriber's credit account is in order, or special arrangements will have to be made to ensure payment. It will be seen that 'pay-per-view' can only be adopted on systems where there is some form of communication back from the viewer to the head end or intermediate switching centre, so that we are generally speaking about switched-star systems when considering this arrangement. It is not strictly necessary to have interaction along the cable network for 'pay-per-view' to take place, since a simple system could be envisaged where a phone call to the head end would request that a particular programme be sent along the line, and the operator would then check on the credit status of the customer before switching through the appropriate programme.

11.1 CONDITIONAL ACCESS

A more recent term embracing the various methods of pay-television is 'conditional access', i.e. the viewer has access to programmes only when certain conditions have been satisfied. Although the most usual condition will be that money has to be paid, this is by no means the only possible arrangement, and it could be that special programmes for the medical profession or containing lists of confidential information could be made available only to those who type in a special password.

The term 'scrambling' is generally used to describe the process by which the pictures are rendered unwatchable, whilst the additional term 'encryption' is used in more complex systems to cover the way in which various 'key' signals are transmitted along with the picture information to enable the decoder box to decipher the incoming signals, unlocking the system so that the pictures can be unscrambled.

Fig. 11.1 shows the basic principles of how any encryption and key distribution method of scrambling works.

We have already mentioned that scrambling is usually carried out to prevent viewers who have not paid from watching particular programmes, but this is by no means the only reason. Some operators use scrambling as a sensible way of minimising their copyright liabilities. As an example, the European broadcaster of 'Sky Channel', a satellite service intended for reception by cable head ends, encrypts its programmes so that it can tell accurately how many viewers are able to watch its service. This means that copyright payments need to be made for only the limited number of viewers connected to cable systems with the approved decoder, whereas if

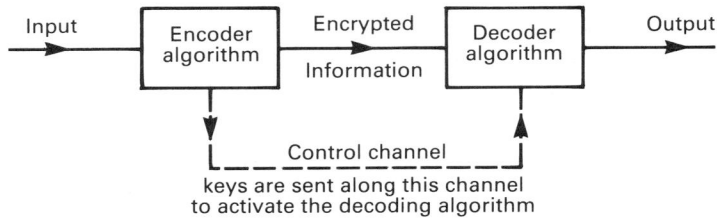

Fig. 11.1 — Basic principals of how any encryption and key distribution scrambling system works.

the programmes were not scrambled, the copyright-protection agencies would be able to claim that payment was due for millions of off-air viewers all over Europe.

Political considerations can also give rise to scrambling, as can be seen by the fact that quite a few European governments, not just those of the Eastern bloc, have insisted that transmissions from satellites are scrambled in order to prevent their own nationals from being able to receive them, perhaps in an attempt to protect their domestic television services from competition.

No practical technique of scrambling, even those using sophisticated encryption techniques, can be considered absolutely secure, and even the complex active line rotation systems described below can be cracked if sufficient time, money and computing power are devoted to the task. Some enthusiasts with considerable technical knowledge will undoubtedly obtain some sort of illicit thrill out of trying to break into the most complex systems, but the cable operator is not too worried about these rare individuals, and is concerned only to have a scrambling system that is secure enough to ensure that the average viewer will not be able to gain unauthorised access. To do this he must use a system which has decoders that are of a design which ensures that it would not be worth while for a 'pirate' manufacturer to build them in quantity, and he should use an encryption system which allows the 'key' information to be changed so frequently that it is just not practicable for a 'pirate' viewer to keep up with the changes.

The scrambling and descrambling of most 'composite' television pictures, i.e. those colour pictures using NTSC, PAL or SECAM, generally gives rise to some loss of picture quality, and care needs to be taken by the operator to ensure that any degradations are kept to an absolute minimum.

11.2 CONDITIONAL-ACCESS TECHNIQUES

The very earliest techniques for preventing unauthorised access to television channels concentrated on removing the audio signals, presumably on the basis that viewers would not watch pictures without the sound, but these days it is always considered essential to 'spoil' the pictures as well. One early method was to insert a second sound carrier to mask the audio information on the main carrier until it was filtered out, and another was to heterodyne the standard audio subcarrier so that it appeared on a higher frequency than the receiver was expecting. The viewer who paid to watch the extra channels would have the necessary frequency downconversion done in this 'descrambler' box.

Such methods are still used to scramble the sound part of a programme in addition to scrambling the video, and various subcarrier frequencies from as low as 12.5 kHz to as high as 62.5 kHz are sometimes used, but other more sophisticated techniques are also in use. Sometimes the conventional sound carrier is omitted, and digital sound signals are sent in the line-synchronising periods using a lower-cost version of the 'sound-in-syncs' type of system used by broadcasters.

11.2.1   Traps

Probably the simplest method of preventing a viewer who has not paid to subscribe to various premium channels from receiving the pictures is to insert a sharply tuned notch filter into the cable system at the point where the customer's service drop is taken from the network, or at some other point on the customer's premises before the receiver. This fairly crude mechanism, usually a completely passive device made up from inductors and capacitors, attenuates the strength of the premium-channel signal so that the receiver does not get enough signal to provide an acceptable picture. Such an arrangement of tuned circuits can be placed in series with normal cable, or, preferably, shunted across the cable to ensure that signals at the frequency of the premium channel do not reach the customer's receiver. Since it would be a relatively simple matter for any would-be 'pirate' customer to disconnect such a device or tamper with its tuning so as to minimise its rejection, traps have to be mounted in a secure environment, such as a sealed metal box. Although it is technically very easy for the cable system operator to remove the trap if a customer wishes to subscribe to the service, and vice versa, the job involves a technician having to visit the premises on each occasion that a change is required, which can be an expensive exercise, especially if the system offers a choice of premium channels on different frequencies. Where several premium channels are offered for a block payment, the frequency response of the trap can be arranged to cover a band of frequencies containing all these programmes.

11.2.2   Reverse traps

If the cable operator inserts an extra radio-frequency 'spoiler' signal into that part of the frequency spectrum between the high-frequency edge of the vision signal and the sound carrier, it distorts both vision and sound signals and generally renders them unwatchable. Customers prepared to pay for the premium channel are supplied with a notch filter which cuts out the interfering carrier and restores the signals to normal. This method suffers from the same practical disadvantages as the normal trap, but in addition it is to be deplored because the addition of extra signals to any cable system is bound to increase the possibility of intermodulation products occurring.

11.2.3   Sync attentuation or suppression

A very common technique is to reduce the amplitude of the sync pulses of the radio-frequency signals by attentuation at the head end. Sometimes the required sync-pulse-timing information (i.e. the encryption information) is then sent as an amplitude-modulated signal along with the sound programme signal, and approved descramblers use this timing information to trigger the generation of new sync pulses which are added to the radio-frequency signals in the descrambler box. More exotic versions of the same technique can vary the amount of sync-pulse attenuation from line to line in a pseudo-random manner, with the encryption signal telling the descrambler how to keep up with the changes in the transmitted waveform.

Such techniques are only partially successful, because many modern receiver designs have synchronising circuits that can keep a picture synchronised even if only a few properly timed synchronising signals are received from time to time, and in any case most television pictures contain enough repetitive information to enable some receivers to synchronise in the absence of perfect sync pulses.

Another rather more effective method of using sync suppression to achieve scrambling is to add a large-amplitude sine wave at line frequency to the baseband video signal. Since the receiver sync-detection circuits will be looking for the most-negative-going part of the demodulated video waveform, this usually being the bottom of the sync pulses, they will lock instead to the most negative part of the sine wave, and thus make nonsense of the transmitted picture. Once again the encryption information, telling the receiver how to compensate for the sine wave which has been added at source, is transmitted in the form of low-level amplitude modulation of the audio carrier (Fig. 11.2(a)).

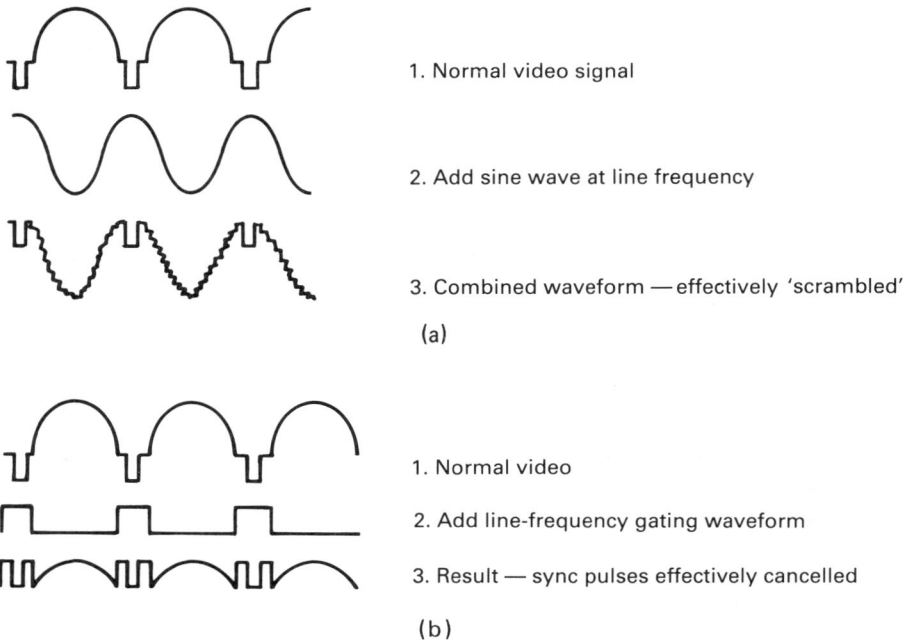

1. Normal video signal

2. Add sine wave at line frequency

3. Combined waveform — effectively 'scrambled'

(a)

1. Normal video

2. Add line-frequency gating waveform

3. Result — sync pulses effectively cancelled

(b)

Fig. 11.2(a) — Showing the operation of the sine wave sync-suppression technique.
Fig. 11.2(b) — Showing the operation of the 'gated-sync' technique.

Yet another method of providing scrambling by modification of the sync puses is to gate-out the sync pulses on transmission by adding a line-frequency gating waveform of opposite polarity to that of the sync pulses, so that they are effectively cancelled out, or significantly reduced in amplitude (Fig. 11.2(b)). The inverse of the gating waveform is then transmitted as amplitude modulation of the audio carrier or sometimes on a completely separate carrier, and when this signal is added to the main carrier signal in the descrambling unit the sync pulses are effectively restored to normal.

11.2.4 Video inversion
At its simplest, this technique merely takes the video signal, inverts it, and uses the decoder box to re-invert the signals for display, but a static method such as this would rapidly be overcome by the addition of a very simple piece of circuitry in the receivers of would-be 'pirate' viewers. More sophisticated methods of this type have therefore been developed, where the video waveform is inverted line by line or frame by frame on a pseudo-random basis, sometimes using scene changes in the transmitted picture to start a sequence of inversion. The information necessary to enable the receiver to follow the inversion sequence can be sent as coded information in the picture-blanking intervals, or can be built into the descrambler unit in the form of a computer program on an IC chip.

Some scrambling systems make life even more difficult for potential pirates by combining various techniques, and a combination of sync suppression and random video inversion can prove very effective.

11.2.5 Variable line delay
This technique introduces various delays into some of the lines of the television picture video signal on a pseudo-random basis, and the encryption signal tells the descrambler to switch in various line delays to correspond with the transmitted picture. Properly controlled, this system can be very effective, giving rise to pictures that are just recognisable but of very poor definition, but unfortunate experiences when the system was first used in France, where pirate decoder designs appeared in amateur radio construction magazines as soon as the new service started, led to the system getting a reputation for poor security. Changes have now been made to the way in which the delays are implemented, which make the system far less open to abuse.

11.2.6 Hidden channels
Some cable systems rely on the fact that they can use frequencies over the cable which are not permitted for over-air broadcast use, and which cannot therefore be received on an ordinary receiver. Thus a cable operator might transmit premium programmes on a frequency below the normal television band, and provide sub-scribers to this service with descrambling boxes that are just radio-frequency convertors which will provide an output at a frequency suitable for use with a normal television receiver. The risk with this method of operation is that viewers may be able to get hold of multi-band receivers which could tune directly to the cable frequency, but in many countries this would be a difficult and expensive business for the would-be 'pirate' viewer.

11.2.7 Scrambling multiplexed analogue component signals — active line rotation
We have seen earlier how a MAC-type signal is made up of individual components which are transmitted in time-division multiplexed form. It has been found that it is

much simpler to scramble and descramble a signal of this type without distortion occurring than to carry out the same processes on the more usual composite signals (Fig. 11.3).

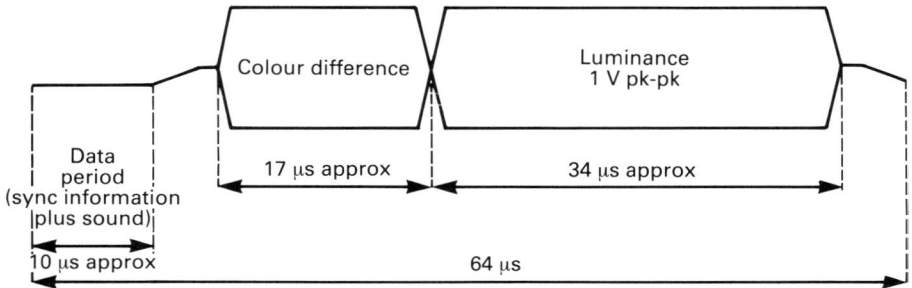

Fig. 11.3 — Basic MAC line waveform.

The technique used is known as *Active Component Rotation,* and Fig. 11.4 indicates how the basic scrambling process is carried out.

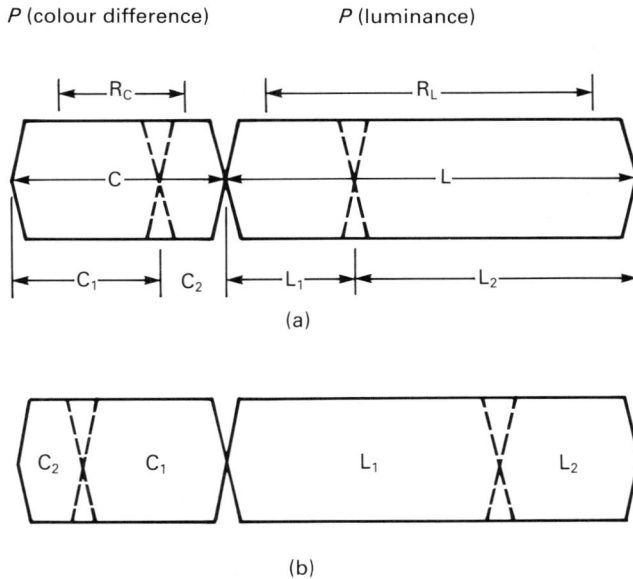

Fig. 11.4 — Simplied diagrams of (a) non-scambled waveform showing possible cut points in the case of double-cut component rotation and (b) waveform scrambled by double-cut component rotation.

The luminance and colour-difference components on each line of the picture are each cut into two parts, the cut points being determined as part of the encryption mechanism by a pseudo-random-number generator. Each of the two parts is then interchanged (effectively, rotated about the cut point) so that the chrominance and luminance information on each line is scrambled before being transmitted. There is a choice of 256 cut points during each of the luminance and chrominance transmission periods.

When both luminance and chrominance signals are cut and transposed, as described above, the technique is known as 'double-cut component rotation', or 'double-cut active line rotation', but the system allows for the simpler option of just the luminance component being scrambled, in which case the technique is called 'single-cut. . . '. The double-cut system is the most secure method, but provision has been made for the simpler method as well because of fears that there might be problems when the double-cut scrambled signals are passed over less-than-perfect cable networks which could introduce amplitude/frequency-response distortion and line tilt. These faults could lead to problems of transients occurring at cut points when the scrambled parts of the picture come to be reassembled in the receiver, and theoretical work suggested that the line tilt would have to be less than about 0.5% if the components were to be put together without error when using a double-cut system.

Practical work on a real cable system [1] showed that any transient effects due to possible problems with line tilt affecting the scrambling system were masked by the small amounts of noise present in any such cable network, and no significant difference could be detected between the single-cut and double-cut forms of scrambling.

Since the use of multiplexed analogue component signals in broadcasting is fairly new, and since the systems now being developed are expected to have to cope with whatever further developments come along in cable, satellite and terrestrial broadcasting in the foreseeable future, it is perhaps not surprising that the conditional-access facilities that have been provided as an integral part of the MAC specification [2] are extremely comprehensive, and, for that reason, rather complex.

A simplified description of the conditional-access system can be understood from Fig. 11.5 and description below, but readers needing more details should consult refs [2] and [3].

It will be seen that the top half of Fig. 11.5 resembles closely our earlier diagram (Fig. 11.1) which showed in generalised form how any encryption and key distribution system works.

The luminance and chrominance components of the MAC signal are first of all scrambled as described earlier, cut points being selected in a pseudo-random binary sequence (p.r.b.s.). The actual point in the p.r.b.s. which is chosen at any instant is determined by a sequence of data known as a 'control word', which can be transmitted over the broadcast system, whether via cable or over-air, in order to synchronise the decoding circuitry in the descrambling unit of the receiver.

Since it could be difficult to introduce scrambling to new or existing services that had not already made suitable technical provisions, many operators think that it would be wise to scramble all MAC signals, whether or not it is intended to charge premium payments for the particular service. If the service is to be free, then a predetermined 'local' control word is used as the encryption key, and a receiver which sees an incoming signal encrypted with this 'local' control word will descram-

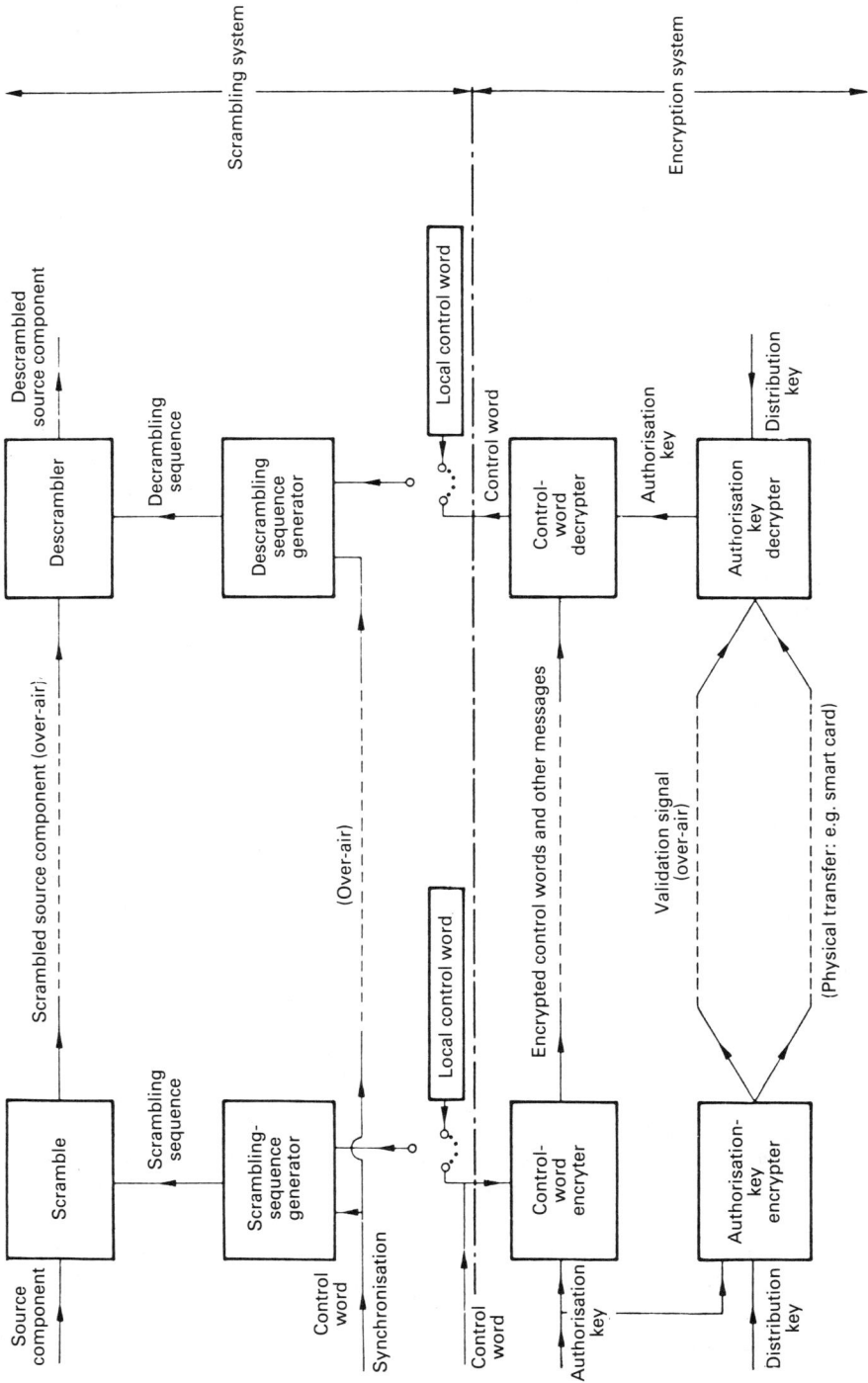

Fig. 11.5 — Block diagram of simplified MAC conditional-access system (Courtesy EBU).

ble the signal without futher ado, so that the viewer will be able to watch the programmes without payment. If at some time in the future the operator decides that the particular programme is to be charged for, he then has the ability to change the control word, so rendering the programme unwatchable until the viewer takes steps to arrange payment for the programme. The idea of scrambling all programmes, free or not, therefore gives the operator tremendous flexibility to cope with future needs.

Where a programme is to be charged for, the operator encrypts the control word by means of an 'authorisation key', and as a further security measure this authorisation key is itself encrypted by a 'distribution key'.

The user cannot see a descrambled picture until his receiver is provided with the correct control word, and he will only be able to obtain this control word if he can decrypt the authorisation key which was used at source to encrypt the control word. To be able to decrypt the authorisation key, the user needs his own 'personal distribution key', which might be built into the receiver or be a personal identification number that he types into the receiver, and also a 'validation signal', which is transmitted along with the programme. The various access control signals, i.e. the encrypted version of the control word and the validation signal, are transmitted in the data burst of the MAC signal and are therefore readily available as an integral part of the MAC transmissions.

This 'built-in' scrambling ability of MAC transmissions is proving very popular with a wide range of programme providers, since the fact that the MAC chip-sets in receivers will already include descrambling facilities means that no separate, expensive descrambling box is required. For this reason many distribution satellite transmissions intended for reception at cable head ends use the MAC transmission format, and the other benefits of using MAC, such as better noise performance and lack of cross-colour, are regarded as an almost incidental bonus. A group of European semiconductor manufacturers has designed the multi-standard MAC receiver chip-set outlined below, which should enable the benefits of a complex scrambling and conditional-access system to be made available to satellite and cable system operators at a very low cost to the viewer (Fig. 11.6).

11.2.8 Public-key cryptography

Various methods of providing the necessary 'keys' can be envisaged, but a system known as 'public-key cryptography' seems likely to gain widespread recognition among the European companies who are broadcasting via satellites to cable head ends [4]. The basic idea is shown in Fig. 11.7, from which it will be seen that the encrytion key, used at the transmitting end, can be different from the decryption key used in the receiver. This means that one of the keys, the transmission key, could be made public, giving a great deal of flexibility to the programme-service provider, so long as the other key cannot be deduced from the public key. The decryption key could be embedded in a 'tamper-proof' silicon chip in the receiver, or could be read from a 'smart' credit card containing a microchip, which the viewer would plug into his receiver.

The matched pair of keys is usually derived from two large prime numbers which are multiplied together, only the product of these being published. The basic security of such systems relies on the fact that it is extremely time consuming to factorise large numbers in order to determine the individual keys, and, as an example, a 312–bit

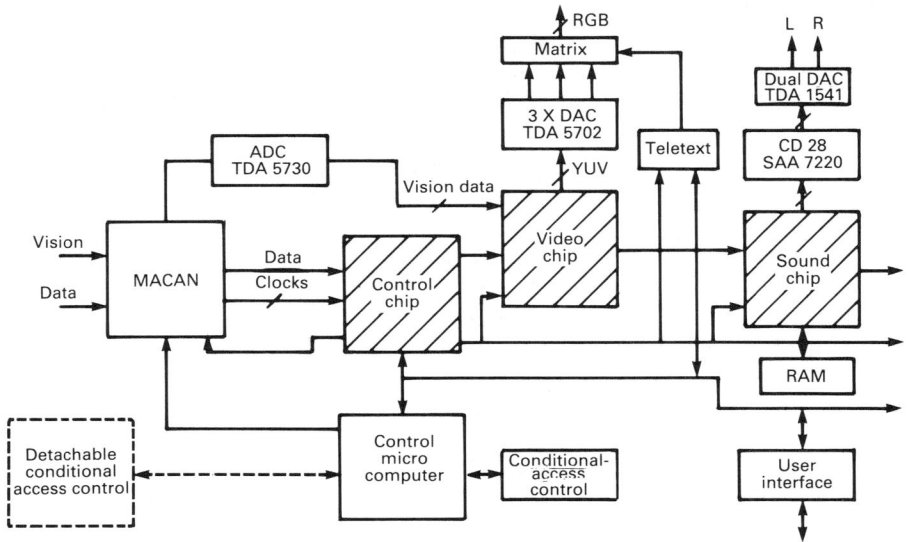

Fig. 11.6 — MULTIMAC decoder chip-set.

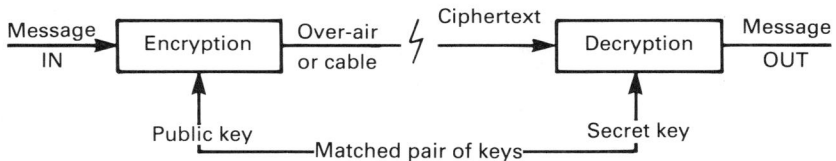

Fig. 11.7 — Basic 'Public-key' system.

number may well take over 20 years to factorise, using the fastest computers available today. No system is 'crack proof', however, and the systems current being considered for use on cable systems have to take into account security, complexity and practicability, and it is interesting to note that system designers are also thinking of building a 'fall-back' or 'recovery' position into their security systems that they are not left totally out of control if the system should be broken into. Experience in the United States with the 'Videocipher' system, where pirates went to considerable technical lengths to extract keys from existing authorised receivers in order to duplicate these for use in 'pirate' receivers, has left system operators in no doubt that their security systems must remain flexible if they are to outwit the illegal users.

REFERENCES

[1] O'Neill and Avon, The distribution of C-MAC in Cable Systems, IBC, 1986.
[2] EBU Tech. 3258 Specification of the MAC/packet family of systems.
[3] H. Mertens and D. Wood, Standards proposed by EBU for satellite broadcasting and cable distribution EBU, 1985.
[4] A. G. Mason, The theory of RSA public key cryptography, IBA internal report, 1987.

12

Cable rules and regulations — a tangled situation

A major problem with trying to decipher the rules and regulations about cable television is that almost every country that has cable has its own ideas about how that cable should be controlled, and in addition it is not surprising that different controls, or a lack of them, often apply to the technical standards and to the programmes that are carried.

Technical standards are probably the simplest to clarify, because the fundamental framework for technical performance and testing is laid down in IEC document 728 [1], a publication that is absolutely indispensable to anyone who needs to know about the technical side of cabled distribution systems, and one that we have referred to frequently during the course of this book. The requirements of this standard have been internationally agreed, and are recognised by countries all over the world.

As well as performance requirements, IEC 728 covers measurement procedures for noise and various forms of distortion, limits for radiation from cable systems and immunity against unwanted signals being induced into the system. It gives clear definitions of the terminology used in cabled television systems and spells out safety requirements.

Individual countries have their own technical regulations as well, British Standard 6513 [2] being a typical example, but the international drive for technical standardisation is fortunately giving rise to a situation where many of the individual standards are becoming 'harmonised'. This will mean that many of the actual words of individual countries' standards may be the same as the words that eventually appear in the IEC standard, once all the stages of negotiation have been completed.

As another example, in West Germany the PTT, the Bundespost, is responsible for laying down technical specifications for cable head ends, and until recently it also had a monopoly on the installation of satellite dishes. This meant that most installations had to be built to extremely 'professional' technical standards, which tended to make the costs rather higher than in some other parts of Europe. This monopoly has now ended, and cable operators can install their own dishes to rather more modest specifications, although again the standards are set by the Bundespost.

In the United States of America, the Federal Communications Commission (FCC) is the regulatory authority, and although cable television in the USA historically grew up in the absence of virtually any technical rules, since the late 1960s the FCC has laid down detailed technical requirements which vary according to the type of cable system under consideration. Four distinct classes of cable system are defined by the FCC:

Class 1: Cable systems which carry standard television broadcasts intended for public consumption.

Class 2: Cablecasting. Cable systems carrying programmes originated by the cable company.

Class 3: Systems carrying programmes that cannot normally be received on a standard television receiver. Conditional-access programmes, subscription services and 'commercial' data services are included in Class 3.

Class 4: Systems including an 'upstream' capability, the so-called interactive services.

12.1 INTRODUCTION

FCC rules lay down such things as the bandwidths to be used and whether or not cable casting and interactive services must be provided, together with requirements about time-signal transmissions, etc. The intention is to ensure that new or extended cable systems make use of forward-looking technology that will ensure that such systems are capable of providing the new services that will arise in the future. Some of the regulations laid down by the FCC can be seen as restricting the powers of the cable operator, or as protecting the subscriber, depending upon which side you happen to be on. As an example, the scrambling of signals in Class 1 systems or channels is strictly prohibited by an FCC regulation; this can lead to difficulties for an operator who wishes to use scrambling in order to reduce energy peaks during the transmission of synchronising signals in order to reduce distortion thresholds on multi-channel systems. Also very much involved in making progress towards common standards for cable systems, especially in the field of interfaces and connectors, are the National Cable Television Association and the Electronic Industries Association. They are particularly concerned to ensure that videocassette recorders and standard television receivers can easily be connected to wideband cabled distribution systems, and are currently trying to agree on a standard for attaching CATV decoders directly to VCRs and television receivers.

The British government employed similar forward-looking thinking to that of the FCC with its insistence that any newly installed tree and branch systems should have the capability of being upgraded to switched-star systems in the future, when it is hoped that market conditions will prove favourable.

12.1.1 Radiation from wired systems

Great care must be taken in the design, and especially in the layout and construction, of cabled systems in order to prevent radiation from the systems causing any form of interference to other authorised users of the radio spectrum, whether these be mobile radio, aeronautical, or broadcast users. In an ideal world the amount of interference to a radio or television receiver operating in the same frequency band as a nearby cabled distribution system would be less than the minimum field strength needed for the receiver's normal operation, but this is extremely difficult to define accurately, since the amount of interference will depend upon the distance of the receiver from the cabled system, and on the frequency in use. IEC Standard 728–1 does not yet contain details of standards for radiation levels for complete systems, saying that they are 'under consideration', but it does give full details of acceptable standards of radiation from individual components of a cabled system, together with details of how the appropriate measurements are to be made on various components. The standard does, however, give details of a method for continuously monitoring radiation from a system, by injecting a specially identified carrier signal into the network, and having monitoring vehicles fitted with equipment designed to receive this particular signal. This method gives no objective or quantitative measurements, but does allow any untoward changes in the radiation of this signal from the cable system to be noticed. This technique works on the assumption that any fault in a cabled system which causes undue radiation will result in a similar effect at all the frequencies used in the system, even though it is recognised that the levels of power radiated will probably vary at different frequencies.

In general, radiation is likely to be worse from cabled systems which are slung from poles than from those where the cables are buried underground, but even underground systems have to come up into distribution boxes, and radiation from above-ground equipment can sometimes be a problem. The weak points as far as unwanted radiation are concerned are usually found to be in the poor screening of amplifiers, splitters and other ancilliary equipment, and it is usually found that the screening of the actual cables, both in trunk and in branch feeders, is quite adequate. Although IEC 728 is as yet of no real help, the line taken by the regulatory authority in the UK, the Department of Trade and Industry, is quite clear, and will probably represent the basis of any future IEC decisions. Basically, the UK authority takes the view that cabled systems have no right to use any part of the radio spectrum, and therefore any radiation must be at the lowest possible levels. This is all very well in theory, but in practice it is realised that some radiation will leak from cabled systems, and so the effects of this have been minimised by choosing the frequencies used within the cable with great care. In general this meant that cabled distribution systems were restricted to using the broadcasting bands, using frequencies not used by local transmitters, and this worked reasonably well, especially as most UK cabled systems carried only a small number of programme channels. The changes that took place in 1984, whereby VHF Bands I and III were withdrawn from broadcasting in the UK, although still used for this purpose in the rest of Europe, led to a new situation where VHF cable systems were no longer carrying all their channels within the off-air broadcast spectrum; this could potentially cause interference to the new mobile communications systems which have been allocated spectrum space in the VHF bands. It was, however, considered impracticable to insist that long-extant VHF cable systems be immediately brought up to the highest of modern standards applying to unwanted radiation, and so the general approach has been to take all possible steps to ensure that interference to the new users of the spectrum is minimised, and new users are also being urged to use equipment that is as immune as possible to interference.

It is worth noting that although optical-fibre cables do not radiate, the opto-electronic devices at each end of the fibres can, and practical problems with the optical modulators is some systems have led to the need for the roadside cabinets containing electro-optical equipment to be screened.

12.1.2 Immunity to external fields

The required level of immunity of cabled distribution systems to radio-frequency fields is defined in IEC 728-1, as is the approved method of measurement, although some engineers feel that it is difficult to measure the immunity of a system in its field environment, since this would sometimes necessitate the setting up of radio-frequency test transmissions of a very high field strengths. A figure of merit is derived for each system, this being defined as the ratio of the wanted signal to the maximum level of the external signal picked up through the system as measured at a system outlet. It is obviously possible to improve this figure of merit by increasing the levels of the wanted signals at the system outlets, but these must not exceed the maximum permitted carrier levels shown in IEC 728–1.8.32, generally 83 dB relative to $1\,\mu V$, expressed as the r.m.s. voltage of the vision carrier at the peak of the modulation envelope, when all outlets are terminated in $75\,\Omega$.

Setting the required levels of immunity is one thing, legislating for cable systems to meet these levels is quite another. A study carried out for the UK Department of Trade and Industry showed that the field strengths from off-air transmitters of various types varied enormously in different frequency bands and in different locations, their effects depending to a large extent on the distance of a particular transmitter from a particular cabled system. To take account of almost all these variations, the study showed that within a typical system, a level of signal immunity within the range 80–130 dB would be required [3]. It was decided that it would not be economically sensible to insist on all cable systems having to meet the upper level of immunity, and that it would be ridiculous to legislate for a lower value of immunity which would not always give interference-free results. In view of all this, the UK regulatory authorities have so far decided not to provide specific regulations for immunity, although any cabled system operator will need to ensure that the performance of his system is acceptable to his subscribers in this respect, as in all others, under the general terms of his operating licence. The Telecommunications Act does, however, give the authorities the power to set specific standards, and this could therefore happen in the future.

12.2 'MUST-CARRY' RULES

We have seen already that the rules governing cabled distribution systems vary tremendously from country to country, but if there is any one rule that is near to being universal, it is the so-called 'must-carry' rule which is invoked by many governments to ensure that the existing public broadcasting services must be carried by the cable system in addition to any other channels that the system operator desires to offer to his customers. In the UK, for example, a 'must-carry' rule applies not only to all existing broadcast programmes from the BBC and the Independent Television companies, but also insists that future direct broadcast services from satellites will also have to be provided on all cable systems. It is interesting that this regulation was vigorously opposed by the CTA (Cable Television Association), the trade association representing many cable system operators, which feels that operators should be able to choose which satellite services they provide. In Finland the 'must-carry' rule goes even further, with cable operators being required to carry not only the appropriate regional services of YLE, the national broadcasting company of Finland, but also any public-service announcements that are intended for nationwide viewing. In Sweden, the cable operator has additionally to provide a local access channel which groups such as churches, youth organisations or even local newspapers can use. In the USA the FCC has recently bowed to a legal attack from cable operators, who managed to convince the courts that the strict 'must carry' rules, were a violation of their rights under the First Amendment.

12.3   A RANDOM ROUND-UP OF REGULATIONS

The law under which cabled distribution services in the UK are controlled is the Cable and Broadcasting Act 1984 [4]. This set up a regulatory body known as the Cable Authority which has the responsibility to license cable operators to provide a service for a specified period, with a maximum of 15 years, in a given area, and the duty to ensure that the detailed provisions of the act are carried out. The Cable Authority has drawn up codes of practice for programme standards and for advertising, and reflecting the public-service nature of broadcasting in the UK, licensed cable system operators must consider such things as the range and diversity of their programmes, the amount of imported programme material, and the provision of programmes which would offend against good taste or decency or which might incite people to crime or disorder.

In contrast to these rules, the cable companies in Austria are not allowed to make their own programmes, but may carry signals received from virtually any local or international station. In Belgium, which with some 80% of all houses now cabled must be the most densely cabled country in the world, cable operators can even carry telecommunications traffic in addition to a wide range of television programmes from neighbouring countries. Denmark is more strictly regulated, with the PTT (posts and telecommunications authority) permitting only telephone companies and local authority groups to operate cable systems, whereas Finland is the most lightly regulated country, with virtually no restrictions upon who may operate a cable system or provide 'cablecasting'-type programming on it. This has led to some 50 different cable television networks being set up in the country, most of them privately owned, and the press and various telephone companies own and operate several of the networks. It is interesting to compare this arrangement with the British system which imposes a duty on the Cable Authority to ensure that no television or radio contractor nor the proprietor of a local newspaper may become a holder of a licence to operate a cabled distribution system [5].

Although Finland has few concerns about who may run cable systems, it does have strict regulations aimed at promoting 'free speech and democracy, supporting national, regional and local culture, promoting debate and adherence to good manners'. In addition, Finnish operators must comply with a unique requirement that obliges them to put any unused 'spare' channels at the disposal of any other licensed cable operator — surely the most effective method possible of ensuring that cable operators make the fullest possible use of their systems [6].

The idea of one operator being compelled to provide another with assistance of some sort recently took a knock in the UK, when the PTT operator, British Telecom, absolutely refused to allow a West London cable television operator to share its ducts under the city streets. These particular ducts serve several huge tower blocks on local-authority land, and it might have seemed reasonable to allow the cable service provider to use these ducts to simplify the problems of getting cables to the hundreds of potential subscribers involved. The dispute was put to the regulatory body concerned, the Office of Telecommunications OFTEL, but OFTEL said that unless the current laws are changed there is no way that BT can be made to share its capital assets. Paradoxically, this is in spite of the fact that the cable operator is obliged under the terms of his operating licence to investigate the use of existing ducts before installing new ones. The Cable Authority has made it plain that it feels that these BT restrictions are unnecessary, since it is known that many such ducts are currently underused, and the Cable Television Association (CTA), which looks after the interests of those companies involved with cable television, has gone on record as saying that cable operators should be allowed to use BT ducts as a right, subject to appropriate financial agreements being reached. In support of its case, the CTA claims that such a right would make the most efficient use of the nation's resources, since it would be wasteful to have to provide new ducts where adequate capacity already exists. It is plain that such a restriction on the use of BT's ducts will slow down the potential growth of cable distribution systems in the UK, and this is a particularly important factor in a country where local planning restrictions generally prevent the slinging of cables from poles in urban areas, a system which is common in many other parts of the world, including the USA.

As might be expected, the French cable systems, which are almost completely funded by the government, are controlled technically and for programming purposes by the PTT, although the day-to-day running of the systems is done by mixed private/public companies known as *sociétés d'économie mixtes*.

In West Germany, the PTT, the Bundespost, provides the cable trunk network, but private operators provide the programme services, and the connections from the network to the customers' homes are provided by different private installers, a situation which can lead to much confusion. As we have mentioned already, the Bundespost sets the technical standards which cable operators must adhere to.

12.4 ADVERTISING RULES

Just as advertising on television is subject to different rules in almost all countries, so advertising regulations vary enormously in their impact on cable transmission systems in different countries. We are familiar with the fact that satellite broadcasting renders it virtually impossible to stop viewers receiving broadcasts across national frontiers, but we should also remember that many European countries are close enough together for cable distribution systems to cross national boundaries, and this has led to many difficulties. Generally speaking, those countries which have strict controls on the amount of content of advertising on their local television or radio services are unhappy about the prospects of unrestricted advertising coming in on the television programmes from across their borders. In contrast the countries taking the more liberal approach to advertising tend to view any restrictions by their neighbours as unfair restraints upon trade, and these differing approaches have led the European Economic Community (EEC) to prepare a green paper (a discussion document) which recommends the harmonisation of broadcasting regulations, including those affecting advertising, throughout the EEC. This document, 'Television without Frontiers' [7], recommended such things as advertisements restricted to blocks between programmes of no more than nine minutes per hour, with programme interruptions for advertisements, the so-called 'natural breaks', being forbidden. Many member governments have expressed strong reservations about the draft proposals, and it appears that it will be especially difficult to reach agreement on matters of copyright and production quotas. Progress towards an accord has been so slow that the EEC draft directive may be overtaken and replaced by another document, a convention which is being drawn up by the Council of Europe. Such a legally binding convention would have some advantages over an EEC directive, but although it would have the status of law in each of the countries concerned, this would not apply until the convention had been formally ratified by each state, which could take years to achieve.

The UK, for example, is fairly typical in its equivocal approach to the proposed convention. It claims to support the principle of the free flow of programmes throughout Europe, but has reservations about having to adopt other countries' regulations on advertising, and would prefer to leave such details out of any agreement.

Similarly, the Belgians claim to support the 'Television without Frontiers' document, but want special restrictions on advertising to be included. Denmark, which at present has no advertising on its domestic television services and therefore normally forbids it on cable channels, is currently having to think hard about the implications of allowing cable operators to receive programmes via satellites, which has been recently permitted, when such services could contain advertisements.

A similar problem faces the Dutch government, which already has severe restrictions on advertising, including a total ban on advertisements on Sundays. The Dutch government imposed on cable operators the duty to ensure that any imported (i.e. via satellite) channels do not carry advertisements aimed at Dutch audiences. As well as being very difficult to do much about, this rule has brought the Netherlands government to the attention of the European Commission, which says that the Dutch regulations are breaking the law, since the Treaty of Rome states that such retrictions discriminate unfairly against transmissions from other member states of the EEC. The Swedish government has similar rules preventing imported programmes from carrying advertising directed at Swedes, and they are likely to come under pressure from the European Commission to liberalise the advertising rules. Austria, too, supports such restrictions on advertising directed at its nationals from other countries.

The Norwegian cable operators are currently not allowed to show advertisements, reflecting the law as it applies to the broadcast services, but the possibility of permitting cable operators to carry pay-television services is under review at the present time. In West Germany, the regional structure of broadcasting makes life even more complicated than usual for the cable operator, and in some areas, such as Dortmund, advertising is forbidden, whereas in others, such as Berlin, advertising is allowed in breaks between programmes.

The selection of rules provided above shows that there is still a long way to go in Europe before all the problems of cross-frontier cable and satellite broadcasting are sorted out, and it is to be hoped that the EEC 'harmonisation' processes will be agreed before too long, since such differences between the practices of various states can only restrict the expansion of cabled distribution services.

12.5 COPYRIGHT

We have mentioned only briefly the problems of copyright, which can currently prevent programmes being re-transmitted over a cable or satellite network unless every single member of the cast of a production is in agreement. Although it may seem to be a *'primâ facie'* breach of a person's rights, the EEC is currently considering whether it would be in the greater public interest if the existing right of a copyright holder to prohibit transmission of a programme in which he has appeared should be removed. This would mean that if agreement to transmit could not be obtained via the normal negotiating machinery, it would be possible for a cable or satellite operator to obtain a statutory licence which would permit transmission and allow a standard fee to be given to the copyright holder. It will be realised that any such scheme would be very controversial, and the initial reactions to this suggestion from the EEC were very favourable from those with cable interests, but copyright holders, including many of the largest broadcasters, were completely opposed to the idea.

REFERENCES

[1] Cabled Distribution Systems, IEC Standard 728–1,1986.
[2] Wideband Cabled Distribution Systems, British Standard 6513.
[3] W. K. Ritchie, An Integrated Services Network. B.T.R.L.
[4] Cable and Broadcasting Act 1984, Chapter 46.
[5] Cable and Broadcasting Act 1984, Chapter 46, Part 1.8.(2).
[6] Finnish Cable Act, January 1987.
[7] Television without Frontiers.

13

The future of cable — all or nothing, or just a long time coming?

The wide variation in the amount of penetration which cabled distribution systems have achieved in different countries of the world makes any predictions as to the way future developments will take place even more difficult than this type of technical stargazing usually is. There are two conventional responses when media-aware people are asked what the future of cable holds. Some envisage a future where all the radio, television and interactive services are brought into every house in the same way as electricity or water; others look at the capability of a single satellite to serve tens or hundreds of millions of people with a wide range of services and say that the lower cost per head of providing satellite services must win out in the end, and that satellites will therefore replace cable.

The cognoscenti shake their heads and say that satellites can never provide the interactive services that the future will demand, but we have considered this point in an earlier part of the book and found that even a humble copper-wire telephone line can provide virtually all the two-way capabilities that it is likely anyone will need for home or small-business use. Even the electronic keys to permit viewers to watch scrambled programmes can these days be sent over air in the form of coded teletext-like signals addressed to individual receivers, so that we certainly do not require a cable to provide the advantages of pay-television.

13.1 AMERICAN PRECEDENTS — A CABLED FUTURE FOR EUROPE?

History has generally shown that what happens in the United States in the field of radio and television happens in Europe a few years later, although recently this has become less true, with some developments, such as the take up of teletext and the home video-recorder, growing faster in Europe and Japan than in the USA. The history of the American usage of cable has very much involved the cable operators in using distribution satellites as a means of keeping cable customers happy by providing them with a wide range of programmes obtained from satellite sources. This close linkage between cable and satellite is already being repeated in Europe, where cable operators who have been restricted for years to relaying only the basic national programmes plus one or two others have now found that the ability to offer a much wider choice of viewing by putting satellite-receiving dishes at their head ends is giving the cabled services a competitive edge over the off-air services.

In countries like Belgium, where over 80% of all television homes are now cabled, the way forward can be clearly seen, and cable operators have the technical potential to introduce more and more new services at a financial cost which will prove attractive to both operator and subscriber. These may involve specialised pay-television channels, interactive services, or even telecommunications services, and the main problems which the cable service providers will have are political and regulatory, since any technical difficulties are usually soluble, and the cost of supplying viewers with new services is generally realistic. The strong existing cable base in many countries of western Europe could provide an exciting future for cabled distribution systems, since cross-border links present no technical problems, and it is relatively inexpensive to provide viewers with an extremely wide range of pro-grammes from different countries. Since the cabled viewer does not need to buy any extra equipment, he may well be prepared to pay extra for a wider choice of viewing, and this money will then be available for reinvestment in programme making, which is really the seed corn of the whole television business. Even more importantly, the ability to watch programmes from other European neighbours may well also have a leavening effect on the closer harmonisation of the various countries of Europe, so that for the first time, Europeans really will come to consider themselves as members of one great entity.

Once 80% of a country like Belgium is covered by cable services it is usually a relatively simple matter to increase this coverage by extensions to the existing networks, although a law of diminishing returns exists, and it may well be that the economics of laying cable to every isolated hamlet mean that it is impracticable to serve all potential viewers in this way. A satellite can provide coverage of much closer to 100% of any scattered population than a cable or terrestrial transmitter network but with a capital cost of around a hundred million pounds for a high-power statellite, this is no economical way to achieve coverage of the final 5% of a population which is already mostly served by cable or off-air services.

13.2   UK CABLE — HARD TIMES AHEAD IN THE SHORT TERM?

Much depends upon the coverage requirements placed upon the broadcasters by the government, and although many countries would consider that a coverage of perhaps 95% of the population would be adequate, the UK government placed broadcasters in a 'Catch 22' situation where they are required to serve 'as many people as may from time to time be practicable' [1]. Thus although terrestrial television transmitters currently cover about 99.5% of the UK population, the job of building relay stations to serve the remaining population continues, and the current situation is that transmitters will be built at the rate of about 25 a year until virtually all groups of 200 or more people are provided with four television channels, even though this has long been a strictly uneconomic undertaking. With such a background, where public-service broadcasting traditions provide almost everyone with a choice of four off-air programme channels for no more than the cost of the receiving licence it is perhaps not surprising that cable has found it hard to compete, expecially since the government's treatment of capital expenditure on cable services for tax purposes has proved far from generous. Under a million homes out of a total of over twenty million in the UK are passed by cable systems, and although accurate figures for penetration, i.e. homes connected as a percentage of homes passed, are difficult to obtain, it is generally agreed that an average figure of about 17% is correct, although some of the newer systems can claim well over 20%.

With such a low baseline it is difficult to see how cable television in the UK can ever make the breakthrough needed for it to become the main means of television distribution, and the promise of three more television channels from a high-powered satellite which can be received on cheap domestic equipment by the end of 1989 will make life even more difficult for the cable operator, although he will, of course, be compelled to carry these extra channels. The root of the problem is financial. Although broadcasters would deny that there has been any element of subsidy in the building of the transmitter network, half the cost has come from the BBC licence fees, and the other half has come from the profits of the commercial television companies, so that the viewer has paid directly or indirectly, but almost without realising it, for the construction of the transmitter network. Since it is not generally permitted to hang distribution cables from poles in the UK, for environmental reasons, all new cables have to be embedded in the ground, and the cost of these ground operations is therefore high; figures of between four hundred and seven hundred pounds per house have been quoted for the cost of each additional home connected, depending upon the location, and it is therefore no surprise that new companies have found some difficulties in raising the necessary capital. How long it will take to make a worthwhile profit on such a huge investment is the major question that would-be investors ask, and when they are told that it may be as long as seven years, many people decide to entrust their savings to a less-risky venture. Fortunately, though, there are those who believe that cable has a future, and in those areas where the latest technologies and the widest choice of programme and interactive services are being provided, there are signs that the viewers appreciate the new services, and that they are prepared to pay for more. Many of these 'new' cable distribution services are in fact extensions and refurbishments of the old cable services which were kept operating by the old Rediffusion company, which kept faith with cable for a period of over 30 years despite the fact that it looked as though the off-air broadcasting services were being given all the breaks.

13.2.1   UK cable — fast forward into the future? — one man's view
In the short term, then, life looks difficult for the cabled distribution systems in the UK, with other technologies seeming to have the upper hand, but as in all crystal-ball gazing, much depends upon the person staring deep into the mists of the crystal. The scenario that follows can be no more than one man's view, but this particular viewer believes that cabled systems in the UK may have a great future in the longer term.

It is 1988, UK viewers have four programme services and the promises of three more from cheap satellite receivers in 1989. The broadcasters are beginning to replace their 20-year-old terrestrial transmitters with new, more efficient ones that will last another 20 years. This will cost tens of millions of pounds, which, once spent, will maintain the 'status quo' until well into the twenty-first century. Cabled distribution systems can only expand slowly against this background, but new satellite services will continue to appear, and will prove popular since they can be received on existing equipment.

Now let us jump forward 20 years. The broadcasters will need to replace their transmitters again, this time at a cost of perhaps hundreds of millions of pounds, unless inflation really can be persuaded to stand still. In order to buy this new equipment, the broadcasters will have to discuss where the money is coming from, and it will not be at all surprising if by that time some satellite operator says he can provide nationwide multi-channel television coverage for a fraction of the cost of renewing the terrestrial network. Although there are some worries that a satellite service will not be as secure in times of national emergency as would be a network of terrestrial transmitters, the financial arguments win the day, and plans are drawn up for the national services, BBC1, BBC2 and Channel 4, say, to be immediately transferred to satellite. The ITV companies protest that they provide regional broadcasting, and that satellite beams are too large and all-embracing to make their form of broadcasting viable from satellites. By the time 20 years have passed, however, the technology will have advanced enough for us to be able to carry much larger, and therefore more directional aerials on broadcasting satellites, so that it will be possible to have a satellite footprint that covers just Yorkshire, or just Scotland, so satellite broadcasting could replace all our current terrestrial broadcast output.

Satellite transmitters still cost a great deal of money, however, and at about the time that this major national investment will be required, the national telecommunications operator, whether British Telecom (BT) or some other body yet to be created, may well have just completed its programme of replacing existing copper-wire telephone cables throughout the land by fibre-optic cables, so that virtually every house has a small fibre-optic cable instead of the previous telephone wires. This programme has taken several years to complete, but has been entirely self-financing, because BT has been able to sell the copper cables it has withdrawn from service at a price considerably higher than it has had to pay for the replacement fibre-optic equipment, which has itself become much cheaper since opto-electronic components have become available in IC chip form. BT has had to plan the replacement programme carefully over a number of years, because it has so much copper available in its ducts that to put it all on the market at once would lower the price of copper in the world's commodities markets. Once the project is complete, however, with over 95% of all homes having a fibre-optic cable input, BT can suggest to the government that it be allowed to carry all existing broadcast services on its fibre-optic network, in order to reach virtually every home in the kingdom. Because it has already paid for the capital costs of the work of installing this huge fibre-optic network, and made a profit out of the deal, it can offer to carry the television services free of charge, knowing that it will be able to reap its profits from the multiplicity of new services that the optic-cable network will be able to carry in the future. The government is unable to resist such a wonderful financial deal, adds a few more 'musts' to its 'must-carry' regulations in order to show that it is still in charge, and gives the go-ahead.

Thus in one mighty leap, cable has overtaken both terrestrial and satellite broadcasting for home use, since who will buy satellite-receiving equipment when all the programmes are already coming into the house by cable? Only the needs of the portable user have not been satisfied by the cable, and by that time there are sufficient satellite services available for the viewer with his brief-case portable satellite receiver with built-in flat-plate aerial to have access to a wide range of other programmes, including the 24-hour international news programmes, and world-wide paging services.

Although most of the present-day cable operators do not yet appear to have taken the possibility of such a scenario on board, BT certainly has, and details have been published [2] of BT's plans for the provision of a national multi-purpose fibre-optics grid. This complex cable network would provide for interactive services, high-quality television pictures, data-transfer services, videotext services, and even picture-phone services.

One extra factor, which could upset some of the careful calculations that have been made by both cable and satellite operators, is the possible introduction of terrestrial microwave broadcasting within a period of perhaps five years. At the present time there is a great deal of interest, especially in the USSR, but also in France and West Germany, in the use of 12GHz for this purpose, and in the USA a system known as MMDS (Multichannel microwave distribution system) is already in use at frequencies of 2.1–2.9GHz, providing customers with up to 10 extra television channels. The Eire government is currently considering the introduction of MMDS, although it is usually referred to as MVDS (microwave video distribution system). Moving even higher up the frequency spectrum, British Telecom is trying out a 29GHz MMDS system, which rejoices in the initials M^3VDS (millimetre wave multichannel multipoint video distribution system) at Saxmundham in Suffolk. The wide bandwidths available at these frequencies could allow for up to 20 extra television channels to be radiated, or for high-definition television signals to be carried in the future. Terrestrial transmitters could serve population groups of a few thousand at a cost very much less than that of cabling. Such signals are affected by rain, need a clear line-of-sight between receiving and transmitting aerials, and travel only a couple of Kilometres under normal circumstances, but this last could prove an advantage, since the same frequencies can be re-used by other operators only a short distance away, without any interference being caused. Frequencies of around 29GHz need only small receiving dish aerials, perhaps as little as 15 cms in diameter, and although suitable microwave receivers would be extremely expensive at the present time, manufacturers believe that the mass production of monolithic microwave integrated circuit chips could bring prices down to sensible levels within about five years.

Although MMDS looks like a formidable competitor for cable, many cable operators are keen to be given permission to use MMDS, because it would allow them to reach a greater number of viewers more quickly than by cable methods, and some say that they would use MMDS only as a 'temporary' method of generating revenue from areas which they have not yet got round to cabling. It seems to this author, at least, that the problems of changing viewers over from their microwave dishes to a cable system at some unspecified future date could represent a major snag for this scenario, although it is true that only the cable system will be able to offer real interactivity. Be that as it may, in the UK the Cable Authority is supporting this line of reasoning, and has welcomed a recent study of the viability of MMDS.

During the latter years of the twentieth century, higher-definition television

(HDTV) pictures which can be shown with excellent quality on large screens of perhaps a metre diagonal will become available on videodisc and perhaps even from special wide-band satellite transmission systems. If cable distribution systems are to keep up with this new technology they will have to make provision for the increased bandwidths that such services will require. Whether HDTV finally comes in the form currently proposed by the Japanese [3], requiring a bandwidth of over 20 MHz, or in one of the enhanced multiplexed analogue components formats requiring about 16 MHz that most European countries seem to prefer [4], cable operators will need to ensure that they have the capacity and the flexibility to deal with the increased bandwidths that will certainly be required. Only by planning to cope with such demands well in advance can operators be sure that their systems will prove adequate, and it is noticeable that one of the claims for the BT fibre-optic system mentioned earlier is that it will be able to deliver HDTV to the home.

Although the picture painted in the paragraphs above shows a rosy future for cabled distribution systems, observant readers will have noticed that it is BT, or some other nationwide operator, which ends up in total control of the nationwide network, a situation which, although familiar in many countries where the PTT runs everything, is totally at odds with what most cable operators would want. There is, however, no really good reason why local operators should not be allowed to provide their own competing services, whilst using the network that BT has laid down, and it is to be hoped that a suitable legislative and control framework can be worked out to permit us to have the best of both worlds.

We end on an optimistic note, therefore, with thoughts of a nationwide fibre-optic cabled system that provides all the services that we could possibly need, including the feature that only cable can provide, a really local service of radio and television programmes and information that can be used to bind together the communities in which we live — true community broadcasting. The only slight cloud on the horizon of our cable landscape is the fact that the horizon is still some distance away; cable in the UK has a great future, but the future may be a long time coming!

REFERENCES

[1] Broadcasting Act.
[2] D. Tatham, Cable '85, Proceedings, OFTEL.
[3] HDTV system parameters.
[4] Enhanced MAC.

Index